AF616820

ENVIRONMENTAL PROBLEMS OF COASTAL AREAS IN INDIA

ABOUT THE EDITOR

Prof. Vinod K. Sharma is an environmental engineer with Ph.D. degree from Indian Institute of Technology Bombay. Currently he is working as Professor and Head in the Department of Postgraduate Studies and Research at the SNDT Women's University, Mumbai (India). He started his career as Civil Engineer and subsequently worked as a Lecturer in the Board of Technical Education, Government of Rajasthan, Assistant Professor of Civil and Environmental Engineering at Jai Narayan Vyas University, Jodhpur and Associate Professor of Environment at Indira Gandhi Institute of Development Research, Mumbai. He has more than seventeen years of work experience that includes research, consultancy and teaching in the area of Environment and Development.

Prof. Sharma has published several research articles, in peer reviewed International and Indian journals, books and monographs. He has organised many academic events and also presented papers in India/abroad. His academic assignments abroad include an invitation from the United States Asia Environmental Partnership to conduct research at the World Bank, USA and an award from the Science and Technology Agency of Japan to work at National Institute of Japan Environment Agency. ***Prof. Sharma*** has completed several research and consultancy projects sponsored by the International organisations such as the UNDP, UNCTAD, UN-ESCAP, CREED, the World Bank and Japan Environment Agency. His current areas of interest include both local and global environmental issues with particular focus on technology and policy aspects of air, water and solid waste pollution, climate change, forests and biodiversity.

ENVIRONMENTAL PROBLEMS OF COASTAL AREAS IN INDIA

Editor

Vinod K. Sharma
Professor and Head
Department of Postgraduate Studies and Research
SNDT Women's University
Mumbai (India)

BOOKWELL, DELHI

First Edition 2000

ISBN: 81-85040-34-6

Published by

BOOKWELL

Sales Office

24/4800, Ansari Road,
Darya Ganj,
New Delhi-110002.
Ph: 91-11-3268786, 3257264

Head Office:

2/72, Nirankari Colony,
Delhi-110009
Ph: 91-11-7251283
FAX: 91-11-3281315
E-mail: bkwell@nde.vsnl.net.in

Printed in India by:
DP's Impressive Impressions,
New Delhi-110059

PREFACE

The coastal areas of India accommodate about one fourth of country's population that is dependent to a large extent on marine resources. Nine of the Indian States namely, Gujarat, Maharashtra, Goa, Karnataka, Kerala, Tamilnadu, Andhra Pradesh, Orissa and West Bengal form the long coastline. In addition, some of the Union Territories such as Pondicherry and Daman and group of Islands including Andaman and Nicobar near the East Coast and Lakshadweep on the West Coast also form the coastal eco-systems of great economic and ecological importance in the country.

Increasing human activities in coastal regions, mainly for livelihood, industrial and other development activities are threatening marine resources. Manifestations of over exploitation and degradation due to over pollution are visible in many parts of the country. The concerns are more serious in coastal cities like Mumbai, Calcutta and Chennai where rising population, rapid industrialisation and unplanned growth result in host of local ecological and environmental problems.

Also, global issues of climate change and rise in temperature put a question mark on the long-term sustainability of the coastal cities. Therefore, immediate measures are required to preserve invaluable marine resources and ensure sustainable development of coastal areas and cities.

This volume deals with some of the important environmental and socio-economic issues related to coastal areas as perceived by various researchers in different coastal states of India. The volume is divided into twenty chapters. Each chapter highlights the problems and issues of concern and suggests some remedial measures to tackle them. It is expected that the volume would be of immense help in understanding the coastal issues in India. In addition to researchers, teachers and students working on any aspect of coastal areas, the volume is expected to be useful for decision-makers and contribute to the government policy for these areas. Comments and suggestions from the readers are most welcome and would be incorporated in the next edition of this book.

-Vinod K. Sharma

ACKNOWLEDGEMENTS

This book contains invited chapters written by distinguished authors working in various coastal states of India. Some of the papers are the part of proceedings of a Seminar on "Coastal Cities in India: Responding to Environmental and Socio-Economic Issues of Concern" sponsored by the International Federation of Institutes for Advanced Study (IFIAS) and International Development Research Centre (IDRC), Canada and organised by the editor at Indira Indira Gandhi Institute of Development Research (IGIDR), Mumbai. I express my sincere thanks to Professor Peter Timmerman of Institute for Environmental Studies, Univeristy of Toronto and IFIAS for his efforts in hosting this seminar at IGIDR.

I am highly grateful to all authors for their timely contribution to this volume. Dr. Kirit Parikh, Director of IGIDR, Dr. U. C. Mishra, Director of Bhabha Atomic Research Centre, Mr. V. K. Phatak of Mumbai Metropolitan Region Development Authority, Mr. G. N. Warade of Department of Environment, Government of Maharashtra and Mr. Debi Goenka of Society for Clean Environment had an extremely useful discussion and gave valuable suggestions during the seminar.

Special thanks to Dr. Jyotsna Maurya of Deccan College Postgraduate and Research Institute, Pune, Ms. G. S. Haripriya and Mr. Santanu Gupta of IGIDR and Mr. Kisley Kishore of Central Institute of Fisheries Education, Mumbai. Help in typing and formatting of the manuscript was provided by Mr Mahesh Mohan of IGIDR. Also, I am thankful to all of them who directly or indirectly contributed towards the preparation of this volume.

-Vinod K. Sharma

AFFILIATION OF THE AUTHORS

S. Ayyappan
Central Institute of Freshwater Aquacultuare,
Kausalyaganga, Bhubaneswar - 751002 (Orissa).

Ramachandra Bhatta
University of Agricultural Sciences,
Dept. of Fisheries Economics,
College of Fisheries, Mangalore - 575 002, (Karnataka).

S. Bhattacharyya
Institute of Wetland Management and Ecological Design,
Salt Lake City, Calcutta - 700 091 (West Bengal).

Anil Baran Bhunia
Central Pollution Control Board,
Shahadara, Delhi - 110032.

Sharad B. Chaphekar
1, Lakshminiketan,
14, Dhas Wadi, Thakurdwar, Mumbai 400 002 (Maharashtra).

Sanjay Deshmukh
Bombay Natural History Society ,
Hornbill House, Mumbai - 400 023 (Maharashtra).

Vidyadhar Deshpande
Urban Development Department,
Government of Maharashtra,
Mantralaya, Mumbai - 400 001 (Maharashtra).

Nishikant Dighe
Bombay Natural History Society ,
Hornbill House, Mumbai - 400 023 (Maharashtra).

Gautam N. C.
National Remote Sensing Agency,
Balanagar, Hyderabad- 500 037 (Andhra Pradesh).

J. P. George
Central Institute of Fisheries Education,
Versova, Mumbai- 400 061 (Maharashtra).

K. Govindan
Regional Centre,
National Institute of Oceanography,
Andheri (W), Mumbai - 400 061 (Maharashtra).

Debanjan Gupta
West Bengal Pollution Control Board,
Calcutta – 700 017 (West Benagal).

Tapas Kr. Gupta
West Bengal Pollution Control Board,
Calcutta – 700 017 (West Benagal).

Arun B. Inamdar
Centre for Studies in Resources Engineering,
Indian Institute of Technology, Mumbai - 400 076 (Maharashtra).

Madhavi Inamdar,
Bombay Natural History Society ,
Hornbill House, Mumbai - 400 023 (Maharashtra).

Ranjana Jaiswal
C-28, 2/2/4/III, Palm Beach Co-op. Hsg. Soc.,
Nerul, Navi Mumbai – 400706 (Maharashtra).

Jayanthi S. C.
National Remote Sensing Agency,
Balanagar, Hyderabad- 500 037 (Andhra Pradesh).

J. K. Jena
Central Institute of Freshwater Aquacultuare,
Kaysalyaganga,
Bhubaneswar – 751 002 (Orissa).

K. Kishore
Central Institute of Fisheries of Education,
Versova, Mumbai- 400 061 (Maharashtra).

A. Kumar
National Botanical Research Institute,
Lucknow, India

Ashwini Kumar
Green Eminent Consultants,
20-Purnima Society,
Fatehgunj, Vadodara-2, (Gujarat).

Rakesh Kumar
National Environmental Engineering Research Institute,
Mumbai – 400018 (Maharashtra).

J. S. Mani
Ocean Engg. Centre,
Indian Institute of Technology Madras,
Chennai - 600036. (Tamil Nadu)

Antonio Mascarenhas
National Institute of Oceanography,
Dona Paula – 403004 (Goa)

J. Nanda
Institute of Wetland Management & Ecological Design,
Salt Lake City, Calcutta -700 091 (West Bengal).

S. Nayak
Space Applications Centre,
Ahmedabad - 380 053, (Gujarat).

D. A. Patil
National Environmental Engineering Research Institute,
Mumbai – 400 018 (Maharashtra).

V. K. Phatak
Planning Division,
Mumbai Metropolitan Region Development Authority,
Mumbai - 400 051 (Maharashtra).

Vinod K. Sharma
Indira Gandhi Institute of Development Research,
Mumbai-400 065 (Maharashtra).

Vijaya Subramanian
C/O N.P. Subramanian,
The Industrial Credit and Investment Corporation of India Ltd.,
Mahalaxmi, Mumbai-400034 (Maharashtra).

O. Sudhakar
Central Institute of Fisheries of Education,
Versova, Mumbai- 400 061 (Maharashtra).

Peter Timmerman
Institute for Environmental Studies,
University of Toronto and International Federation of
Institutes for Advanced Study,
Toronto, Canada.

Piyush Tiwari
Housing Development Finance Corporation,
Mumbai -400 020(Maharashtra).

Ravi Shankar G.
National Remote Sensing Agency,
Balanagar, Hyderabad- 500 037 (Andhra Pradesh).

Raghavswamy V.
National Remote Sensing Agency,
Balanagar, Hyderabad- 500 037 (Andhra Pradesh).

Vikas Tondwalkar
Planning Division,
Mumbai Metropolitan Region Development Authority,
Mumbai – 400 051 (Maharashtra).

G. K. Tripathy
Indira Gandhi Institute of Development Research,
Mumbai- 400 065 (Maharashtra).

M.D. Zingde
Regional Centre,
National Institute of Oceanography,
Andheri(W), Mumbai - 400 061 (Maharashtra).

TABLE OF CONTENTS

ISSUES IN COASTAL AREAS WITH SPECIAL REFERENCE TO COASTAL CITIES

Vinod K. Sharma and Peter Timmerman

INTRODUCTION

From the beginning of history, people have settled near water bodies, and cities have mostly grown up along river banks or on coastlines. This was due to the availability of water and other basic needs such as free sewage disposal, easy transportation, and strategic concerns. Economic activities were concentrated in coastal regions because trade facilitated more business opportunities. Increased economic and employment opportunities led to immigration and the mushrooming of population. This has resulted in ever growing environmental and socio-economic problems in coastal zones, and increasingly in coastal cities.

All over the world, people are moving to the cities and to the coasts. By the year 2000-01, it is estimated that over 50% of the world's population will be living in cities, and about 50% of that population will be in coastal cities. Thus, coastal cities are under increasing stress, but they are also very often the engines of economic growth and creative innovation in each country.

Recently, coastal cities have been identified as being among the most vulnerable regions for the potential impacts of climate change. The conjunction of local and global environmental issues, even without further reference to other problems, indicate that coastal areas, particularly coastal megacities demand more of our attention. Yet the relationship between coastal zone management in general, and the special concerns and priorities of coastal cities in those coastal zones is not well studied; and coastal cities in particular are not well thought through economically, politically, and ecologically.

Economically and politically, coastal cities suffer from jurisdictional overlaps and conflicting demands. Ecologically, these cities are the convergence point for two intensely complex ecosystems: the natural ecosystem of the coastal zone, and the dynamic urban ecosystem of the city. But this makes them very difficult to study and to manage. For example, while there has been extensive work over the last twenty years in both the management of urban areas, and in coastal zone management, there has been little research specifically linking these two together into what can be identified as "coastal cities" research.

Environmental and socio-economic problems of coastal ecosystems in developing country like India encompass a large number of anthropogenic activities. Three of the four megacities in India viz. Mumbai, Calcutta and Chennai are on coasts. In adition, there are several coastal towns playing crucial role in the economic development of the country. Some of the major problems of all these urban and coastal eco-systems include discharge of domestic and industrial effluents, agricultural wastes, radioactive and thermal wastes; tourism and shipping; oil spills and over exploitation of the living coastal resources etc. Human activities create stress on the ecosystem beyond its tolerance limit that can pose hazards to the coastal and marine environment, and to the health and safety of the population living in the coastal areas. A city like Mumbai is a living example of the both the terrible problems and the immense opportunities provided by this global phenomenon.

The rest of this Chapter provides a general overview of some of the problems and issues of the coastal areas of India, with special reference to the burgeoning coastal cities.

COASTAL STATES OF INDIA

Nine of the Indian states, some union territories and groups of islands are the coastal areas of ecological and economic importance in the country. In addition to their importance for livelihood of the people living in these areas, they are also strategic locations for industrial development. Several industrial establishments cause pollution pressure on these areas. (See Figure 1.1)

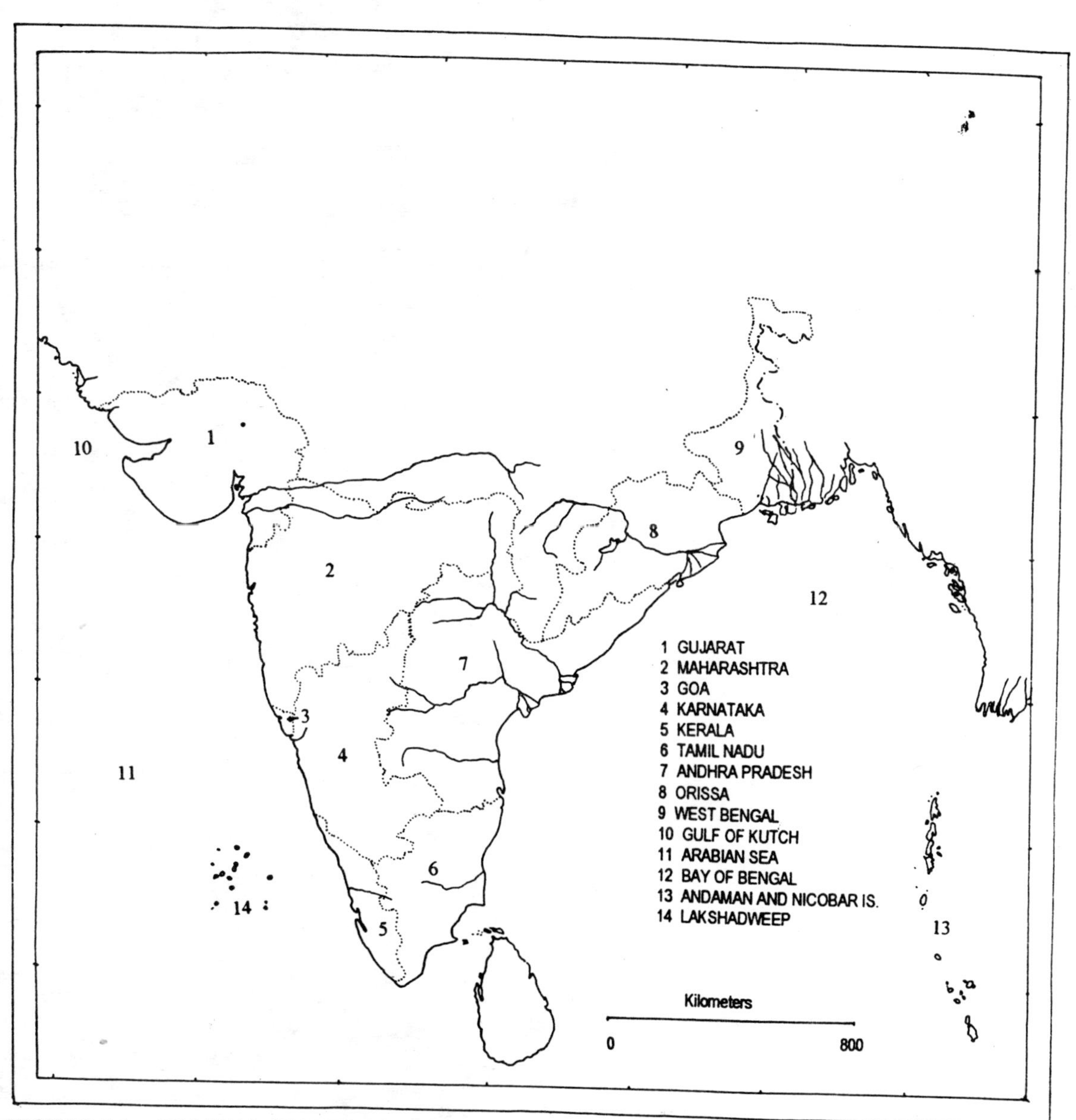

FIGURE 1.1 – COASTAL STATES OF INDIA

Table 1.1 shows statewise coastal length, area (upto 25 km from shoreline) and major industries housed in these areas as assessed by the Central Pollution Control Board (CPCB) of India. While the state of Gujarat has the longest coastline, the state of Maharashtra tops the number of highly polluting industries situated in the coastal areas. Other important details (in brief) for coastal states of India are given in subsequent paragraphs. Population figures are based upon the 1991 census data.

Table 1.1: Coastal Length, Area and Industries in Coastal States of India

Coastal State	Coastal Length (km)	Area* (km^2)	No.of Industries
Gujarat	1663	41,575	35
Maharashtra	720	18,000	167
Goa	140	3,500	2
Karnataka	290	7,250	3
Kerala	560	14,000	26
Tamilnadu	860	21,500	30
Andhra Pradesh	930	23,250	30
Orissa	450	11,250	4
West Bengal	200	5,000	7

* Upto 25 km from the shore line

Gujarat

With a population of about 41.31 million and a total area of about 196,024 km^2, the state of Gujarat is characterised by two gulfs - the Gulf of Kutch and the Gulf of Cambay. The coastal belt houses rich mangrove forests. Some of the important river outfalls in the state are Sabarmati, Tapi, Narmada and Saraswati. This state is the second largest in terms of industrialisation in the country having textiles, petrochemicals, fertilisers and pesticides, aluminum, refinery, pharmaceuticals, pulp and paper, etc. as major industries. The marine ecosystems in the state have been subjected to over utilisation and increased pollution. As shown in Table 1.1, about 35 industries located at or near the coast discharge their effluents into the coastal waters. In addition, discharge of untreated domestic waste and waste from other coastal activities such as port and harbour, oil refining and exploration, fishing, recreation etc. also put pressure on marine resources. The estimated total waste generation from coastal areas includes about 0.57 million cubic metres per day (MMD) of effluent and about 9500 tonnes per day (TPD) of solid waste.

Maharashtra

The state houses the metropolitan city of Mumbai (Bombay) that is known as financial capital of India. The population and area of the state are about 78.94 million and 307,762 km^2, respectively. The rocky and broken coastal belt, locally known as "Konkan" belt, has several small bays and creeks. The main industrial units include textiles, electronics, chemicals, fertilizers and pesticides, vegetable products and soaps, most of which are concentrated near or around the city of Mumbai. Encroachment due to land use changes required for urban sprawl and industrialization are the major threat to the coastal ecosystems of the state. Since times immemorial, the sea has been utilized as pollution sink for dumping various types of waste. Liquid effluents of about 0.08 MMD and solid waste to the tune of 2628 TPD are generated within the coastal stretches of the state.

Goa

The beach resources of Goa are of international interest and attract millions of tourists every year. This is relatively a small state with population of 1.17 million and area about 3702 km^2 and coastal stretch of 140 km. Mandovi, Zuari and Damanganga are the major river outfalls and contamination of these estuarine ecosystems is of major concern in the state. Pollution loads are generated from tourism and recreation activities, port and harbour, mining activities, household and industries. The liquid effluent of 0.012 MMD and solid waste of 1.4 TPD have been assessed within the coastal areas of the state.

Karnataka

The state of Karnataka is rich in economic minerals such as gold, high-grade iron-ore, Manganese, copper, magnetite, quartz etc. Small rivers like Kalinadi and Swarnanadi form the outfalls and their deltas have dense forest including mangrove fringes. With a population of over 44.98 million the state occupies 191,791 km^2. Some 0.04 MMD of effluent and 76 TPD of solid waste from the coastal areas is generated which is mainly from fertilizer and iron-ore processing industries. Fishing and recreation activities also put pressure on coastal resources.

Kerala

With a population of about 29.1 million and a total area of about 38,864 km^2 the state of Kerala is one of the finest states in terms of natural flora and fauna and scenic beauty. Several small rivers flow into the Arabian sea. The state is also rich in traditional and modern fishery processing industries. Besides, the state has strong defense related activities as a strong naval base is situated here. Chemicals, fertilizers and fish processing and other industrial activities combined with domestic, tourism and recreation near coastal stretches generate about 0.15 MMD of effluent and about 2431 TPD of solid waste. Port and defense related activities also form the source of contaminant load in the coastal waters.

Tamilnadu

Cavery is the major east flowing river in the state having main tributaries as Vallar and Coleroon forming deltaic mass at the coast fringed with mangroves. The population of 55.86 million in an area of 130,069 km^2 contributes to increased coastal pollution. Among the industrial establishment, aquaculture, leather, chemicals and fertilizers, cement, mining and automobiles etc. are major source of pollution which add some 0.38 MMD of effluent and 9112 TPD of solid waste. The intensity of aquaculture and seafood processing industry can be estimated with the fact that about 75% of total effluent load comes from this industry.

Andhra Pradesh

Like Karnataka, the state of Andhra Pradesh is also endowed with rich variety of minerals Godavari and Krishna are the major rivers flowing within the state. The state has population of about 66.51 million and an area of 275,068 km^2. A total of 118 industry lying within coastal areas include as much as 88 are aquaculture farms which contribute to more than 80% of effluent load into the sea. In addition to aquaculture, glass, sugar, pharmaceutical etc. are major industries. A heavy effluent load of 2.47 MMD and solid waste load of 7191 TPD has been assessed by the CPCB.

Orissa

Here Subarnarekha, Bramani-Baitarani, and Mahanadi form the major outfalls and are fringed with dense mangrove ecosystems. About 31.66 million people live in an area of 155,782 km^2. There are only four industries within the coastal areas of the state and the comparatively low effluent load (.001 MMD) is added to the sea. However, solid waste generated is around 3505 TPD which includes waste from domestic, commercial and industrial establishments. Most of the industries in the state are mineral based and limestone, dolomite, bauxite, chromite, graphite, china clay etc. are the major mineral found here.

West Bengal

The state of West Bengal has distinct feature of being first in several aspects of coastal resources in India and even in the world. For example, Sunderbans is the largest delta in the world having extensive mangrove forests. The Ganga-Brahmaputra river system transports as much as the one-fifth of the total suspended load carried by the world rivers. The population of the state stands at about 68.08 million and area about 87,853 km^2. Major industries are steel plants, textiles, silk, automobiles, jute, leather, ceramics and glass. The solid waste is comparatively less (only 25 TPD) but effluent load of 0.02 MMD is added to the sea.

COASTAL CITIES IN CRISIS

From the above, we can see that there is a mix of concerns in the coastal zone, ranging from the degradation of deltas to agricultural runoff and others. While not underestimating these concerns, coastal cities in many of these areas represent special "hot spots." If we are unable to solve the problems of rapid urbanization, it is unclear how the rest of the coastal zone problems will be resolved.

If we take Mumbai as an example, which could be multiplied elsewhere, the migration to Mumbai from inland means that almost every year the equivalent of a new city is being added. This puts an extraordinary burden on all facets of city planning and management. In the past century the growth of Mumbai has meant the infilling, building over, and polluting of much of the natural ecosystems - lakes, creeks, rivers - that support life. What remains is also under great stress.

To these stresses we can also add the shadow of changes that are likely to occur in the world's climate over the next hundred years. These will increase the effects of old hazards, and may well create new hazards as yet unknown. Nevertheless, although climate change is not by any means the most serious immediate threat to the ecosystems and the quality of life in the world's coastal cities, we can identify some of the most serious potential impacts as

- an increased risk of flooding and impeded drainage;
- salinization of freshwater supplies;
- higher water tables which may reduce the safety of foundations; and
- beach erosion.

In the face of the importance of these cities and the potential damage that might have to be responded to, it is perhaps surprising to discover that the special concerns and priorities of coastal cities are not well represented economically, politically, and ecologically. This is discussed in more detail in the next chapter entitled "Coastal Cities: A New Agenda."

THE COASTAL EDGE

Despite being under increasing stress coastal zones are increasingly powerful attractors of population, and this is intensified by the drawing power of cities. Coastal cities have made significant contributions to the economic growth and development in most countries around the world. The movement of people to these cities proves that life in the city, with all of its problems, presents opportunities that are at least perceived to be unavailable elsewhere.

However, while the 21st century is going to be the "urban century"; it will also be the century of "economic and environmental sustainability". If the earth is to be inhabitable by 11 billion people, living under the threat of global climate warming, and other environmental threats, we will have to do a great deal better in ensuring sustainability. Learning how to accomplish sustainability is not only a necessity, but will provide substantial economic opportunity.

Coastal cities are pivotal for this, and members of coastal cities are now beginning to speak of the "coastal edge". The coastal edge is the advantage that cities on substantive bodies of water have, both as information and economic centres, and also as innovators in learning how to live together in harmony with the natural world. It is also clear that cities around the world may have more in common with cities elsewhere than with their own hinterlands. There is certainly a widespread cultural convergence among the high elite classes, who are seeking both access to global information and access to high quality environments. Internationally, one can find new connections being made among cities, whether it be at conferences like Habitat II (1996) or through Non-

Government Organizations (NGOs) like the International Union of Local Authorities (IULA), the International Centre for Local Environment Initiatives (ICLEI), and similar others.

These are also signs that cities are increasingly finding that out of necessity and the movement towards the devolution of power to local authorities, they are having to solve their own problems. Coastal cities are increasingly finding that they need to balance a wide variety of civic goals – these include ports, tourism, maintaining a high quality of life for all citizens, and so on. Coastal cities are among the most important incubators of experimentation. They are also "early warning systems" for new problems and new solutions.

THE COASTAL CITIES PROJECT

To highlight the problems of coastal cities and investigate the way for sustainable development of coastal cities, several international agencies are making efforts. In a similar effort, since September 1996, the International Federation of Institutes for Advanced Study (IFIAS) and the International Development Research Centre (IDRC), Canada began to explore the possibility of establishing a network of coastal cities around the world. Among the earliest and most substantive respondents has been the Indira Gandhi Institute for Development Research (IGIDR), a member of IFIAS, with a long-standing concern for Mumbai, for the Indian subcontinent, and for the resolution of global issues generally. A seminar on "Coastal Cities in India: Responding to Environmental and Socio-Economic Issues of Concern" sponsored by IFIAS and IDRC was held at IGIDR in July, 1997.

The seminar was designed as an initial step in the development of a larger initiative to promote the sustainability of Mumbai and other regional coastal cities. The seminar brought together a wide variety of experts from different disciplines and institutions. All of the participants seemed committed to a greater collective understanding of the current situation with regard to Mumbai as a coastal city, and all were generally concerned to assist in the preparation of common responses to the mounting challenges. The goals of the ambitious project of IFIAS/IDRC are:

- to develop a network of coastal cities committed to the sustainable ecomanagement of these cities;
- to develop and sustain multi-client partnerships among researchers, institutes, experts, and potential sponsors of coastal city issues;
- to explore new models of information delivery and "green financing" for coastal city concerns; and
- to utilize the capacities of new media for international networking and support.

The first year of the project has been designed to identify potential partner cities and possible case studies of coastal city issues. In addition, because there has been so little original research done on coastal cities per se, an emphasis has been placed on seminars including appropriate background papers and other research. A useful agenda for near-term research and planning for coastal cities includes:

- A state-of-the-environment (SOE) background/baseline document on natural and urban ecosystem representing facts, issues, and trends and then building on recent SOE reports of coastal cities/areas.
- A proper dissemination and knowledge sharing by organizing seminars, workshops and conferences.
- Use of outcome of above academic events as the basis for a strategic planning process.
- Establishing regional and international links within the coastal cities/areas.

It is expected that this volume would be a first step in attaining some of these goals. This would also form a basic reference document that would help in devising the future research needs and strategic planning for ecomanagement of coastal areas/cities in India. This should not be seen as fundamentally separate from discussions of coastal areas in general: in fact, it is the integration of coastal cities into their coastal zone more sustainably that is the preferred outcome.

2

COASTAL CITIES: A NEW AGENDA

Peter Timmerman

ABSTRACT

This Chapter outlines necessarily selective overview of some of the themes associated with coastal cities. The possibility of further use of the "ecosystem approach" to coastal cities is discussed. Some special problems concerned with coastal cities, specifically with reference to impending climatic change have been described, a discussion of relevant issues is given and conclusions for future research are drawn.

INTRODUCTION

Coastal cities will inevitably be a main focus of concern in the 21st century. Seventeen of the top 25 megacities of the world around the year 2000 will be coastal cities. The migration of population into cities is part of the concurrent movement of people to coastal areas worldwide and coastal cities intensify that drawing power. Historically, cities have been located on coasts and rivers to take advantage of their locational and related ecological benefits.

As mentioned, what is extraordinary is that, given their obvious current and impending importance, coastal cities have been almost completely ignored as a research topic, except as the accidental byproduct of individual case studies of cities that happen to be on the coasts, or as a marginal reference in a study of generic coastal zone impacts.

This is also a function of disjointed mental maps. The conceptual boundary between the coast and city exacerbates the constant difficulty of managing due to overlapping jurisdictions, institutions, and responsibilities. We can add to that the problems involved in relating the dynamics of an artificial or socially created ecosystem ("the city") with the natural dynamics of a coastal ecosystem.

A BRIEF ECO-HISTORICAL SKETCH

A city has been defined as "an open ecosystem for perpetuating urban culture by exchanging and converting great quantities of material and energy". Many of the world's cities from earliest times have been situated near or alongside large bodies of water, including rivers, lakes and oceans to obtain these great quantities of material and energy; and for other reasons such including military advantage and ease of transportation. Coastal locations have possessed substantial health advantages, in that winds and tidal scouring have helped to dilute the accumulation of urban wastes -- solid, liquid, and gaseous. However, the early symbiosis between coastal ecosystems and urban ecosystems usually becomes dysfunctional over time, as the city expands.

The initial shape of the city and its specific orientation to its local space are usually based on the natural constraints and advantages of the site. Over time -- and especially in the modern era -- the city begins to detach itself more and more from the local "natural template", to the point where its appearance and physical structure, as in the suburb or the industrial park, could be from anywhere. The tension between this kind of detached urban form and the continuing need to service it from the local ecosystem with its own characteristics, is found practically everywhere, and nowhere more complexly than in today's coastal cities.

Similarly, as cities become larger and the nature of the human economy changes, and as the impacts of human settlements on coastal ecosystems become steadily greater, the natural ecosystem becomes substantially altered to meet the needs of the city. This alteration may be the natural result of changes in loading patterns and extractions or through deliberate engineering works for everything from transportation to waterworks.

Many coastal cities were established or grew large during the industrial revolution and European expansion. These cities have similar structures, with substantial port facilities, heavy industry, and railway lands that, until very recently, were the sole users of the part of the city closest to the water's edge. With the decline of that kind of use, we consistently find a rediscovery of the waterfront for tourism, housing, and recreation, and the need to redevelop the severely contaminated industrial landscape.

With the rediscovery and rehabilitation of the waterfront, the city begins to reorient itself towards the water, and we often find the beginnings of an environmental awareness of the city as a true "coastal city." The coastal area begins to be seen as a different kind of resource.

It is one of the goals of a coastal city perspective to build on this rediscovery process, and to facilitate the rethinking of the relationship between the city and its coast. We suggest that some of the following characteristics of the emerging city-coast interaction worldwide will appear upon analysis:

1) Cities and megacities around the world are rapidly degrading and simplifying their coastal ecosystems. In some areas, such as Hong Kong and Tokyo, the harbour area is being reduced and even eliminated through contamination, dredging and infill.

2) The ecological reasons why cities were originally located on coastal zones are under threat. Cities can use their surplus wealth to protect the ecosystem from further damage, or to protect themselves temporarily from the consequences of the damage. In the latter case, they often increasingly draw resources from the hinterland, which also becomes degraded, adding to the negative feedback loop. The more degraded the hinterland becomes, the more likely there is to be immigration to the cities, increasing the need for hinterland resources.

3) There is a relationship between the quality of city life and the quality of the local natural ecosystem, but this relationship is complex, and is only expressed directly in modern cities when there is extreme stress (e.g. extensive flooding; sewer system breakdowns; closed beaches).

4) The city ecosystem and the coastal ecosystem will have different, but connected, cycles of operation. The scale relationships (temporal and spatial) between these will also be complex.

5) The social response to the deterioration of the urban ecology will often encourage the private replacement of public services to ensure clean water, refuse collection, electricity, and personal security for individual wealthier households.

6) The policy responses to changes in coast-city interactions will be complicated by multiple jurisdictions, and exacerbated by a bifurcated "mental map" that separates the coastal ecosystem from the city.

ECOSYSTEM MODELS FOR COASTAL CITIES

There are available a number of detailed ecosystemic analyses of natural coastal ecosystems, but far fewer studies of the "city as ecosystem" that can be married to such analyses. There has been very little follow-up application of the methodologies developed through the Man and the Biosphere Programme of UNESCO in the 1970s and 80s. Furthermore, there have been very few attempts to develop the policy implications of the metabolism of a city, or to study urban metabolism in a variety of different physical environments.

It is a contention of this presentation that the opportunity for this kind of approach is now better than it has been, in spite of its past history. There is now a widespread recognition that integrated, multisectoral research and management is both important and necessary for many environmental problems, and that an ecosystem approach provides an appropriate integrator. In at least one jurisdiction -- The Canadian Great Lakes -- there is a legal framework, a research community, and at least 10 years of experience in exploring an ecosystemic approach to the region's problems, including Canadian and American coastal cities. Furthermore, there is a general movement towards Integrated Environmental Assessments as a managerial framework which strongly indicates the possibility of complementary work in ecoystemic analysis.

"Ecomanagement" or ecosystem management is usually applied to natural areas, but has been more broadly extended to take into account the reconceptualization of the "human ecosystem", which includes both the impacts on, and alterations of natural ecosystems by human beings, as well as the special new engineered systems built by human beings for their own purposes. One benefit of an ecosystem approach is that it provides a common ground for diverse visions and jurisdictions in a designated area.

Ecosystem science provides a defensible, cumulative resource for managers and decision-makers. It is particularly appropriate to coastal cities, with their unique mix of built and natural environments. An ecosystem approach is based on a number of working assumptions. One of the Toronto Waterfront Commission Reports, Barret and Kidd (1991) identified some of the following implications of taking an ecosystem approach to an urban area:

- A broad definition of the environment is used (including natural, cultural, physical/built, economic, etc.).

- The ecological boundaries and integrity of the area are primary, and the political boundaries are secondary.
- Attention is focussed on links, relationships, and connections between hitherto obscured elements of the common ecosystem.
- The dynamic nature of ecosystem processes is recognised.
- Concepts such as vulnerability, sustainability, resilence, and carrying capacity are developed as working tools for regional understanding.
- The goals of the approach include the maintenance and restoration of the integrity, productivity, and health of the ecosystem.

To these we can add the following further implications:

- The economic productivity of the region is seen as enhanceable by improvements in ecological sustainability, and not in rivalry with the environment.
- Ecomanagement provides new bridging concepts and methods that can help in crossing traditional jurisdictional and disciplinary divides.
- New information sources, benchmarks, and stakeholders are implicated in adopting an ecomanagement approach.
- Because of the recogniton of potentially dramatic ecological systems changes, ecomanagement implies concern for cumulative impact assessments, and prioritizes the substantive and justifiable margins of safety in the use of ecosystems, as well as the prevention of irreversible damage.
- Consultation and public participation processes are seen as essential to obtaining sustainable political support over the long-term.

The ecosystem approach encourages and often necessitates multiple points of entry for participation. The capacity to understand and monitor the system depends in part on the quality and quantity of local information: this empowers local stakeholders, and legitimates their presence. For example, the World Bank Water and Sanitation Program for Low-Income Urban Populations (PROSANEAR) found in 1994 that involving local community leaders and beneficiaries in basic design choices together with the design engineers have dramatically lowered per capita investment costs and increased the sense of project ownership (which in turn improves maintenance costs).

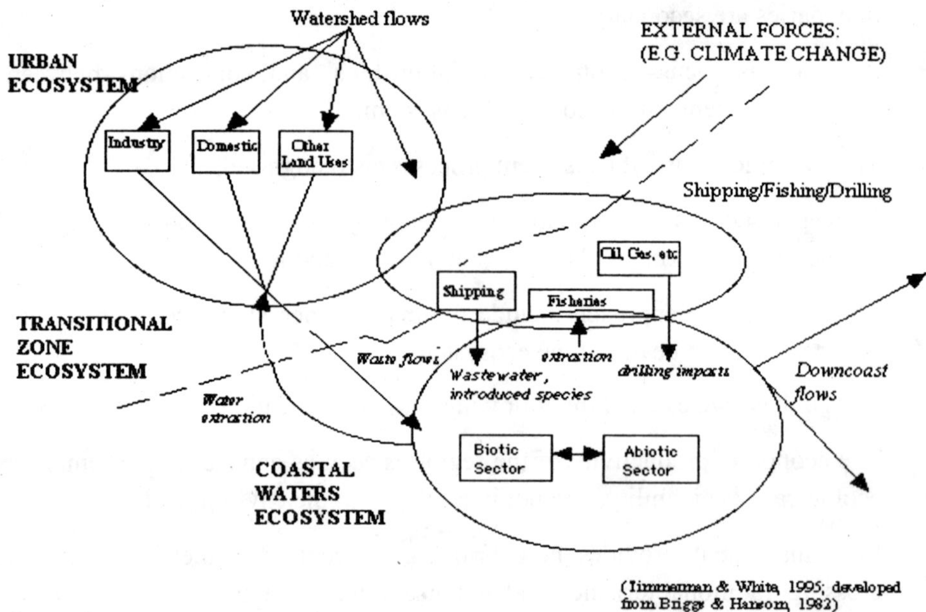

There have been a number of models of how an integrated ecomanagement plan or process should develop. It is important that there is no one model that will fit every case. It must also be recognised that the developing understanding of the complex systemic issues must be linked directly to the evolving policy development and implementation regimes. A generic sequence for ecomanagement would include, as basic steps:

1. Initial inventory assessment and mapping of the potential ecomanagement area, including physical and built environments, as well as identification of relevant stakeholders and institutions.
2. Establishing of administrative and political co-ordination through various institutional mechanisms and participatory processes.
3. Identification of agenda of concern, and initiation of negotiation towards possible common objectives, goals, and responses.
4. Ensuring leadership, resources, and accountability in the implementing process.

5. Promoting continual reassessment of ecomanagement process.
6. Investing in long-term sustainability and widespread ownership of the ecomanagement approach.

SOME IMPLICATIONS OF CLIMATE CHANGE

Climate change is not by any means the most serious immediate threat to the ecosystems and the quality of life in the world's coastal cities, but it is a stress that will exacerbate a range of other problems. Nicholls and others (1995) have argued that the most serious potential impacts from sea level rise are:

1) an increased risk of flooding and impeded drainage;

2) salinization of freshwater supplies;

3) higher water tables which may reduce the safety of foundations; and

4) beach erosion.

Many of the burgeoning cities in the developing world are being forced to spread out onto vulnerable, low-lying land, which also often turns out to be an important wetland regulator of the health of the local ecosystem. Groundwater withdrawals, land infilling, and so on, are parts of this process. What this means is that the "new towns" and shanty villages sprawling out from the core city are especially vulnerable; and of course, they are often where the poor and the landless congregate (though in some cities there may be outrider suburbs or wealthier new towns as well).

Many coastal cities have had long experience of dealing with flooding and storm surges. Cities in the developed world -- and the older cores of cities in the developing world -- have traditionally responded to these threats with substantial protective engineering works ranging from dykes to the raising of houses on pilings. For these coastal cities, the problems in responding to sea level rise will not be from lack of know-how, but from the difficulties of assessing the costs and priorities for protection and adaptation.

Relevant policy responses include:

- the control of the rate of groundwater extraction;

- land-use planning controls and financial dis-incentives (e.g. lack of guaranteed insurance) to exclude buildings from vulnerable locations within an urban area;
- construction and management of infrastructure with the dangers of sea-level rise in mind, including drainage, sewage, solid waste, and transportation networks;
- determining the proper mix of "soft engineering" and "hard engineering" on the coastal zone itself.

Other New Concepts

"Metrofitting" - The rebirth of waste sites and abandoned or underused areas in major cities are a recent theme. The costs of reconnecting old infrastructure are usually far less than building new sewage lines, etc.

Green Infrastructure - The use of greenbelts and watersheds as corridors of influence. The mapping and ecodevelopment of these existing or recent zones provides a visual focus and ecologically appropriate strategy.

Ecological Footprint - What is the actual cost in land and other resources in servicing the city, and the increasing levels of suburbanization? The ecological (and economic) advantages of higher densities in cities can now be more vividly contrasted with the lower densities and higher servicing costs of suburban life styles.

Towards Green Financing

1. It is important for political and economic reasons to be able to capture the costs and benefits of existing and proposed activities as a baseline. If this is not yet done, an environmental audit can be helpful.
2. It is vital to cut down your own overheads. This includes not only internal conservation, but also considering privatization of certain non-essential services.
3. Privatized services should be guided by the results required, not by how you get there, i.e. the methodologies. Government should steer, not row.
4. There is a growing consensus that local governments can charge directly for certain services based on the benefits received, provided the services are not essential, that the "polluter pays" rather than "victim pays", and that income distribution is not worsened.

5. A different operating framework helps identify new partners and alternative sources of funding. An ecosystem perspective provides a scientific justification and a political weapon for making new connections, e.g. into the health sector, into the banking and insurance industries.
6. Horizontal connections between different city experts leads to new pooling of knowledge and resources, and the use of other examples as evidence for one's own community that different approaches can work.
7. Innovative industries focussed on environmental improvement are now signficant technological engines and high-level job creators in many developed countries. A focus on this in Latin America could pay high dividends.
8. "Knowledge brokering" is a new term for engaging NGOs', academics, and other expert researchers in support of decision-makers in the cities. A knowledge broker is an individual or institution that assists in bringing together the most relevant experts and the decision makers to determine immediate and medium term needs, while sustaining the capacity to make higher-level connections and assessments over the longer term.

MAIN ISSUES AND RESEARCH QUESTIONS

The main issues of future concern for coastal cities include:

- preventing or minimizing encroachment onto critical natural ecosystems; encouraging urban renewal through waterfront and shoreline regeneration.
- ensuring amenity values of beaches and waterfront properties; coping with financial constraints and multiple jurisdictions;
- maintaining the "water infrastructure" which includes fresh water supply, sewage, wastewater treatment (if any);
- protecting against local flooding and subsidence through continuing engineering works;
- planning for projected sea-level rise, increased storms, saltwater intrusions, etc., from climate change;
- adjusting to changes in "port city" economies.

Among the research community, relevant research topics include:

- Are there "rapid" or modified environmental audits of coastal cities that can be carried out without vast resources? How can comparative analyses best be carried out, especially between developed and developing country coastal cities?
- What are the most important early indicators of the breakdown of ecological services to the coastal city?
- Can the increasing costs to the hinterland of maintaing the coastal city as it degrades the functonality of its natural support systems be measured, and modified?
- Can the degradation of the hinterland and consequent inmigration to the coastal city be restrained?
- How will the coastal impacts of climate change be locally translated into the already existing array of problems affecting coastal cities?
- How can suburbanization be replaced or supplemented by enhanced city quality?
- How would the introduction of an ecosystemic approach support more appropriate policy responses to changes in coast-city interactions?

REFERENCES

Advisory Committee on Protection of the Sea (ACOPS). 1995. "South East Asian Seminar on the Management of Coastal Cities and Towns" held in Pattaya, Thailand, August 1994.

Bird, Eric C. F. 1993. Submerging Coasts: The Effects of a Rising Sea Level on Coastal Environments. Chichester: John Wiley & Sons.

Boyden, S., S. Millar, K. Newcombe, and B. O'Neill. 1981. The Ecology of a City and Its People: The Case of Hong Kong. Canberra: Australian National University Press.

Briggs, D.J. and J.D. Hansom. "Potential Role of Ecological Mapping in Coastal Zone Management in Europe", Ekistics. Vol. 49, No. 293, March-April 1982.

Delft Hydraulics and the Tidal Waters Division, Rijkswaterstaat. 1993. Sea Level Rise: A Global Vulnerability Assessment. Netherlands: Delft Hydraulics and the Tidal Waters Division, Ministry of Transport, Public Works and Water Management, The Netherlands.

Dix, G.B. 1986. "Alexandria 2005: Planning for the Future of an Historic City", Ekistics, No. 318/9, May/August), pp. 177-186.

Douglas, Ian. 1983. The Urban Environment. London: Edward Arnold.

Francia, C and F. Juhasz. 1993. "The Lagoon of Venice, Italy" in Coastal Zone Management: Selected Case Studies. Paris, France: OECD.

Frassetto, R. (ed.) 1991. Impact of Sea Level Rise on Cities and Regions. Venice: Marsilio Editori.

Hamza, A. 1989. "An Appraisal of Environmental Consequences of Urban Development in Alexandria, Egypt", Environment and Urbanization, Vol. 1, No. 1, (April) pp. 22 - 30.

Hough, M. 1984. City Form and Natural Process: Towards a New Urban Vernacular. London: Croom Helm.

Intergovernmental Panel on Climate Change, Response Strategies Working Group, Coastal Zone Management Subgroup. 1992. Global Climate Change and the Rising Challenge of the Sea. Report of the March 1992 update to IPCC. Ministry of Transport, Public Works, and Watermanagement, The Hague, Netherlands, for the IPCC.

IPCC, Working Group II. 1995. "Coastal Zones and Small Islands" (Chapter 9) in Impacts, Adaptations, and Mitigation of Climate Change: Scientific-Technical Analyses (2nd Assessment Report). Cambridge: University Press.

IUCN. 1993. Cross-Sectoral, Integrated Coastal Area Planning (CICAP): Guidelines and Principles for Coastal Area Development ed. John Pernetta and Danny Elder. (IUCN Marine Conservation and Development Report).

Kingsley, G. T., B.W. Ferguson, B.T. Bower, and S.R. Dice. 1994. Managing Urban Environmental Quality in Asia. Washington, D.C.: World Bank, Technical Paper 220.

Koudstaal, R. 1987. Water Quality Management Plan North Sea: Framework for Analysis. Rotterdam: A.A. Balkema for IFIAS (IFIAS Research Series, Coastal Waters Nr. 1).

Lang. R. 1994. "Urban Ecosystem: From Concept to Application", in Human Society and the Natural World: Perspectives on Sustainable Futures, eds. D.V.J. Bell, R. Keil, G. Wekerle. Toronto: York University, Faculty of Environmental Studies.

Lee, B.J., H.A. Regier, and D.J. Rapport. 1982. "Ten Ecosystem Approaches to the Planning and Management of the Great Lakes", Journal of Great Lakes Research, Vol. 8, No. 3, pp. 505-519.

McCarney, P. 1991. "Draft Terms of Reference for Local Consultants Working on 'World Cities and Environment: A Five City Consultation Process'". Toronto: Centre for Urban and Community Studies.

Milliman, J, J. Broadus, and F. Gable. 1989. "Environmental and economic implications of rising sea level and subsiding deltas: the Nile and Bengal examples," Ambio, vol. 18, No. 6, pp. 340-5.

Nicholls, R.J. 1995. "Coastal Megacities and Climate Change", Geojournal, Vol. 37, No. 3 (Nov).

Nicholls, R.J. and S.P. Leatherman. 1994. "The Implications of Accelerated Sea-Level Rise for Developing Countries: A Discussion", Journal of Coastal Research, Special Issue No. 14 (September 1994).

OECD. Coastal Zone Management: Selected Case Studies. 1993. Paris: OECD.

O'Riordan, T. and Vellinga, P. 1993. "Integrated Coastal Zone Management: The Next Steps", Keynote Speech, World Coast Conference.

Patchamuthu, Illangovan et al. 1995. "Preparing and Implementing an Urban Environmental Management Strategy and Action Plan for Colombo, Sri Lanka" in Urban Management Programme (UNDP), 1995.

Plait, R.H. S.G. Pelczarski, B.K.R. Burbank (eds.). 1987. Cities on the Beach: Management Issues of Developed Coastal Barriers. Chicago, Ill.: Dept. of Geography, Univ. of Chicago.

Rijsberman, F. 1993. "The Effectiveness of the Helsinki Convention as a Tool for the Integrated Coastal Resources Management of the Baltic Sea", Coastal Zone Management: Selected Case Studies. Paris: OECD Documents.

Rolt, L.T.C. 1974. Victorian Engineering. Harmondsworth, Middlesex: Penguin.

Royal Commission on the Future of the Toronto Waterfront. 1992. Regeneration: Toronto's Waterfront and the Sustainable City -- Final Report. Toronto: Queen's Printer of Ontario.

Stren, R. E. and R.R. White. 1989. African Cities in Crisis: Managing Rapid Urban Growth. Boulder, Colo.: Westview Press.

Tolley, M.J. and S. Jelgersma (eds.). 1992. Impacts of Sea-level Rise on European Coastal Lowlands. Oxford, U.K.: Blackwell, for the Institute of British Geographers.

United Nations Environment Programme/World Health Organization. 1992. Urban Air Pollution in Megacities of the World. Oxford: Blackwell.

White, R. R. 1994. Urban Environmental Management: Environmental Change and Urban Design. London: John Wiley & Sons.

World Coast Conference. 1993. Preparing To Meet the Coastal Challenges of the 21st Century: Conference Report. The Hague, Netherlands: Ministry of Transport, Public Works, and Water Management, (RIKZ).

3

PROBLEMS OF THE COASTAL ECOSYSTEMS IN INDIA: A FOCUS ON MUMBAI

Vinod K. Sharma, Piyush Tiwari and Ranjana Jaiswal

ABSTRACT

In this Chapter, Indian coastal ecosystems of major importance have been reviewed briefly. The focus has been on the mangroves, coral reefs, seagrasses and fishery resources of India. Rising population pressure has resulted in serious environmental and socio-economic crises in Indian coastal cities. The impact of major anthropogenic activities on coastal resources is illustrated. Local environmental problems coupled with mismanagement of the coastal resources are threatening these life support systems, particularly in developing countries. Global warming and sea level rise would further aggravate problems of coastal cities. Some of the policy measures are outlined which require further research to make them adaptable for sustainable use of the coastal resources and solving problems of coastal cities in India.

INTRODUCTION

The coastal ecosystems comprise of estuaries, lagoons, mangrove swamps, salt marshes, seagrass meadows and coral reefs. Each ecosystem has distinct characteristics in respect of flora, fauna, and their environment. These organisms and their population are totally dependent on the biotic factors of the marine environment. Any change in their environs, alters the existence of the coastal animals. Coral reefs are the spawning grounds of aquatic life, and seagrass beds are important feeding grounds for fishes, which seek refuge as juveniles in the mangroves. Wetlands serve as a habitat for aquatic animals including migratory birds. Mangroves and coral reefs also act as a buffer against wave and tidal force, stabilizing and safeguarding the shoreline. Seaweed has extensive application in food, textiles, cosmetics, pharmaceutical, fodder and fertilizer industries. Coastal zones have multiple uses including fishing, aquaculture, heritage areas and natural reserves, forestry, navigation, defense, power generation, sand mining, human settlements, disposal of wastes, tourism and recreation. While these activities play a significant role in the economic development of a country, they also have some negative impact on coastal ecosystems.

India has a long coastal belt of about 7500 kms that includes the Bay of Bengal on the east coast and the Arabian sea on the west coast. Nine out of the twenty-five states have their own coast. With economic liberalization, each state is competing for investment and certainly coastal line adds to its advantage because ports can be developed there. The Exclusive Economic Zone (EEZ) extends upto 2.02 million square kms and 14 major rivers, 44 medium rivers and 162 minor rivers discharge about 1645 km^3 of fresh water into the sea, annually. These rivers also carry millions of tonnes of silt alongwith their discharges. Several urban settlements along the coast discharge huge quantity of domestic and industrial effluents into the sea. This has been a practice for a long time without consideration to the scientific studies for assessing the assimilative capacity of the coastal marine environment. The problems which coastal cities like Mumbai, Calcutta and Chennai are facing today, other coastal cities would also face in the future. It becomes important, therefore, to study coastal cities in India.

This study gives an overview of the coastal resources of India. The vulnerability of these resources to over exploitation and degradation due to human activities is highlighted. The study also focuses on the problems and issues of the coastal megacity of Mumbai. Finally, some aspects of an efficient coastal zone management have been mentioned. However, these require substantiation by further detailed studies so as to formulate appropriate policies for sustainable use of coastal resources and eco-management of the coastal cities.

INDIAN COASTAL RESOURCES

Different coastal ecosystems of economic and ecological importance have various utilities for human beings. A general discussion of various coastal resources in these terms and with special focus on India is given below.

1. Mangroves

Mangroves provide shelters and are the nursery grounds for aquatic life. Mangroves are vital natural systems that need to be protected as a near-shore nutrient source, as breeding, feeding and nursery grounds of marine organisms, and for land building, stabilization and protection purposes. These wetlands are being destroyed at an alarming rate for conversion of mangrove swamps into aquaculture ponds and salt pans, land reclamation, for timber, firewood and viscose rayon manufacture, and expansion of human settlements into mangrove areas. Pesticides from sewage and cultivation lands affect the mangroves vigorously. Oil pollution results in the death of mangrove seedlings and associated fauna.

On the Indian coasts, mangroves are found along the islands, major deltas, estuaries and backwaters. There are about 45 species of mangroves, which constitute an important resource in India. Mangroves constitute a significant portion of the coastal wetlands on which a large percentage of coastal population is dependent directly or indirectly. Rapid depletion of these resources due to above mentioned reasons can be estimated by the fact that in 1982-83 the total area of mangrove forests in India was around 681976 ha that was reassessed by Ministry of Environment to be 356500 ha only in 1985(UNEP, 1985; GOI, 1987). Gangetic Sunderbans of West Bengal, Andman-Nicobar Islands, Krishna, Cauveri, Godavari and Mahanadi deltas are some of the best mangrove formations of India as mentioned below. In addition, comparatively small in area and scattered mangroves are found on the west coast.

Andaman and Nicobar Islands

The total mangrove forest estimation of Andaman and Nicobar Islands is 78000 ha, of which 96% occurs in the north, middle, Baratang and south Andamans. Mangrove formations are mainly seen at Kimois (Car Nicobar), Galathea creek (Great Nicobar), and in sheltered bays at Nancowry. Total mangrove cover in Great Nicobar has been estimated to be 2450 ha (Jagtap, 1992). Degradation of mangrove forests along the eastern side of South Andaman near Majeri was observed from satellite data (Krishnamoorthy *et al.,* 1993).

Tamilnadu

Estimated mangrove area in Cauvery Delta is between 500-1000 ha and is available at Pitchavaram, Muthupet and Chatram (Blasco, 1975). Anthropogenic pressure and increase in salinity due to the blocking of fresh water supply from Coleroon (for irrigation purpose), cause declination of mangroves at Pitchavaram. Recent developments in tourism with the construction of cottages within the mangrove forest areas and frequent boating trips for the tourists, even to the core area of the mangroves disturb these ecosystems considerably. Disposal of sewage, and fertilizer and pesticide from nearby cultivation lands cause destruction to mangroves.

Maharashtra

Mangrove forests along the west coast of India was estimated to be 1140 km^2 including 34 species. From satellite data, mangroves in Maharashtra are observed along the creeks of Vijaydurg, Shastri, Vasisthi, Savitri, Dharmatar and Vasai with a total cover of about 21000 ha (Untawale *et al.,* 1982). Mangroves are observed in Raigad and Sindhudurg districts also.

Goa

Distribution of mangrove forest along the Goan coast from aerial photographs, found to occupy an area of 2000 ha, majority of which occur in Zuari (900 ha), Mandovi (700 ha), Cumbarjua (200 ha), Chapora (100 ha), Terekhol (30 ha), Sal (30 ha), Galgibag (20 ha) and Talpona (10 ha) estuaries (Untawale *et al.,* 1982). Intensified siltation by the construction of bunds for agricultural and fish farming causes degradation of mangroves.

Gulf of Kachchh

Here, mangroves occur in association with corals. Multidate satellite data analysis of the mangrove cover of the Gulf of Kachchh showed a decreasing trend from 1975 (276 km^2) to 1986(219 km^2). A net loss of 57 km^2 mangroves over an interval of 11 years was observed. The loss was also evident in the boundaries of Coastal National Park, where a decrease was about 116 km^2 between 1975 (139 km^2) and 1985 (33 km^2). Successive conservation measures from 1985 till 1988 extended the mangrove cover by 22 km^2 (Desai *et al.,* 1992). Enhanced sedimentation was the reason attributed to these losses. Damages due to excess harvesting of mangroves, overgrazing of young mangroves by camels had increased the siltation rate, causing mortality of the adjacent coral reef.

2. Coral Reefs

Coral reefs are among the planet's most diverse ecosystems. One reef can support as many as 3,000 species of marine life. It provides spawning grounds for fishes and other aquatic life. Coral reefs are massive deposits of calcium carbonate and the $CaCO_3$ precipitated by global reef system is about 600 million tonnes per annum (Wafar, 1990). They require a complex environment such as a hard substratum, sediment free water and minimum temperature of 20° C. Reefs grow best at temperature around 25-29° C which explains the occurrence of corals in the tropics and subtropics. The optimum salinity level for coral communities is 30 ppt. Corals are mostly observed in shallow waters less than about 50m depth because of their dependence on intense sunlight.

Live corals play a vital role in maintaining water quality suitable for coastal aquaculture. Efficient nutrient cycling, retention and recycling mechanism sustained in the coral reefs result in high productivity. They are used in carbide industry and white cement industries. They are also used for ornamental and decorative purposes.

Scientific observations indicate an increase in coral diseases, infections due to bacteria, algae and fungi resulting in death of corals. This ecosystem is fast degrading because of the quarrying of corals by industries for raw material, destructive fishing methods, marine pollution, harvesting of coral skeletons and reef animals for ornaments and edible products. Untreated domestic sewage causes eutrophication on corals. Since reef are attached to the substrate, migration to escape from sediment laden water is impossible. Increased sediment depositions, not only kill the corals but also reduce the fish biomass dwelling on it due to their habitat and food loss.

About 342 species of coral reefs in the Indian Ocean include fringing and barrier reefs, sea-level atolls and elevated reefs have been described (UNEP, 1985). In India, coral reef occurrences are observed along the northwest and southeast coasts, Lakshadweep, Andaman and Nicobar Islands. Except Lakshadweep corals, which are atolls, others are fringing reefs.

Andaman-Nicobar and Lakshadweep Islands

Coral reefs of Andaman and Nicobar Islands belong to fringing and patch type of reefs. Heavy land drainage and siltation due to deforestation results in the death of corals around the Islands. Saw dust from timber factories at Chatam and Mayabandar, port activities around Port Blair and coastal pollution increase the mortality of corals. In

Lakshadweep corals 4200 km^2 of coral reefs are reported consisting of around 69 species in Minicoy atolls(WWF, 1992).

Tamilnadu

The Gulf of Mannar located in the southeast coast of India covering approximately 10500 km^2, is a major reef area sheltering wide varieties of fishes, seaweeds, algae, sponges etc. It had been declared as the Coastal Biosphere Reserve under the Wildlife Protection Act, from February 1989. The reefs are extending from Rameshwaram island to Tuticorin in northeast-southwest direction along a chain of 20 Islands, made up of dead corals and surrounded by underwater live corals. Productivity of the fringing corals of the Gulf of Mannar varies from 2 to 5 $gmC/m^2/day$ (MOEF, 1987). Siltation from mainland alongwith quarrying leads to the declination of reef population. Stirring of bottom sediments during northeast monsoon deposits sediments on the reefs causing destruction of the coral colonies.

Gulf of Kachchh

In Gujarat, reefs are found at Mundra, Pirotan Island, Jokhua, Dwaraka and in between the islands in the Gulf of Kachchh. Corals are associated with mangroves, algae, seagrass and fishes. Mining of coralline material about 0.5 million tonnes per annum by Digvijay Cement Company causes major destruction of reefs (Desai *et al.,* 1992). Temporal analysis of satellite data showed the declination in reef area, from 217 km^2 in 1975 to 180 km^2 in 1985 and 123 km^2 in 1986. Thus, a net loss of reef area is 94 km^2 in 11 years. Within the Coastal National Park also, loss of about 64 km^2 reef is observed between 1975 (117 km^2) and 1985 (53 km^2) due to the heavy sediment load. Subsequently, due to the enforcement of conservation measures between 1985 and 1988, reef area had increased by 28 km^2 (Desai et al, 1992).

3. Seagrass Beds

Seagrasses are useful in cleaning the polluted coastal waters and support aquatic life. It is a base for food, fertilizer, drug and pharmaceutical industries. The growth and survival of seaweeds depend on salinity, temperature and sunlight. Seaweed serve as substrate for algae and marine animals, with a very high productivity rate of about 500-1000 $gmC/m^2/yr$. The growth and survival of seagrass depend on salinity, temperature and sunlight. Seagrass flora in India comprising of 12 species mainly occurring along the southeast coast and Lakshadweep Islands.

Andaman and Nicobar Islands

Nicobar group of Islands comprised of 7 species of seagrasses, occurring in Car Nicobar, Great Nicobar, Katchall and Nancowry. Seagrasses were observed in inter-tidal zones and their growth was patchy and mixed due to non-uniform sandy substratum (Jagtap, 1992).

Tamilnadu

The productivity rate of seagrass in the Pitchavaram mangroves ranges from 0.31 to 1.38 gmC/m^2/day. Six genera and eleven species of seagrasses have been observed in the Gulf of Mannar region. Heavy influx of nutrients from sewage on seagrass beds, ensue in eutrophication and hypoxia. Temperature and salinity are the rudimentary factors for this ecosystem. Thermal discharge from power plants and brine run-off from salt pans (Tuticorin) affect the seagrass beds extensively.

Lakshadweep Islands

Seagrass in the lagoons of Lakshadweep and Minicoy Islands were estimated to be about 112 ha and 40 ha, respectively. From aerial photographs, major seagrass beds were observed along the shores of Kadamat, Minicoy, Amini, Kavaratti, Kalpeni and Agatti (Jagtap and Inamdar, 1991).

Maharashtra and Goa

Patches of seagrasses occur in the mangrove swamps at Shirgaon creek in Ratnagiri district of Maharashtra. The extent of segrasses in the Goa state at Terekhol, Chopra, Mandovi and Zuari estuaries are very less.

4. Fishery Resources

With an annual foreign exchange earning of about Rs. 17242 million (one US$=31 Rs.) from the exports in 1993, fish and shellfish industry contributes to the country's economy significantly. Coastal fish production in India was 2.3 million tonnes in 1991, 2.5 million tonnes in 1992 and predicted 2.57 million tonnes for 1993. Over exploitation and wasteful fishing is prevalent on Indian coasts (FSI, 1993; Sharma *et al.*, 1997). Some of the major coastal sites of importance in terms of fisheries are mentioned below.

Andaman and Nicobar Islands

The average fish landing in Andaman and Nicobar Islands during 1984-85 was about 6587 tonnes (Ansari and Abidi, 1989). The total catch rate in the Andaman sea for deep sea lobsters and prawns is 11 kg/hr (Ali *et al.,* 1991). A survey of oceanic tuna resources around Andaman and Nicobar Islands conducted during Oct. 1991-March 1992 indicated that 9-11° latitudes are more productive in respect of Yellowfin tuna followed by sharks (FSI, 1992 & 1993).

Tamilnadu

Annual coastal fish landings of Tamilnadu coast is 244759 tonnes per annum. Prawns followed by crabs, mullets, oysters and other common fishes dominate fishery resources at Pitchavaram. Perches were significant in the catches from the Gulf of Mannar between October 1991 and March 1992, in the depth zone of 200-300 m, a rate of about 1915 kg/hr with shrimp alone about 422 kg/hr was observed(FSI, 1992 & 1993).

Maharashtra

About 19% of the total coastal fish landings of India is contributed by Maharashtra in which the Thane district alone shares about 23.6% (MPEDA, 1987). Demersal resources survey along the Konkan coast during October 1991 to March 1992, recorded the maximum catch rate of 724 kg/hr in the north (March) and 1341 kg/hr in south (December), between 100 and 200 m depth (FSI, 1992 & 1993). The average catch potential observed at Thane, Bassein and Dharmtar creek systems adjoining Mumbai were 0.68 tonne/km^2, 0.22 tonne/km^2 and 0.20 tonne/km^2 respectively (Jyothi and Nair, 1990). Annual primary productivity near Mumbai harbour, at Thal was observed to be 1105 tonnes, with a peak in August. This is due to the high growth of phytoplanktons during monsoons attributed to the relatively low salinity and increased photosynthesis. This acts as fish feeds and results in higher fish production (Varshney *et al.,* 1983).

IMPACT OF ANTHROPOGENIC INTERFERENCE

The coastal ecosystems of India have been endangered by several kinds of human activities along the coasts. Some of them are mentioned as follows.

Overfishing

Over exploitation of fishery resources beyond its regeneration capacity and destructive fishing methods are hazardous to the aquatic life. For example, in 1991, the estimated maximum sustainable yield for the coast of Goa was around 60,000 tonnes per annum but total fish catch noticed for 1995 was 64930 tonnes (Sharma *et al.*, 1997). Bottom trawling for shrimp destroys or disturbs pearl oyster beds and chunk beds, beyond its restoring capacity. Also, trawler fishing near coral colonies stirs up the fine silt and sediments, the accumulation of which onto the reefs jeopardizes the corals. Stake nets and bag nets destroy juvenile fish. In shallow waters the juveniles and larvae may be hit by the propeller blades of mechanized craft. Construction of barrages and salt water barriers to protect coastal agro-forestry alters the ecosystem and sometimes leads to the extinction of living species. Declination in the benthic macrofauna in Coleroon estuary was observed due to the construction of fishing harbour (Ayyakkannu, 1990).

Shipping

For entry of ships into the shallow sea shore, harbours and ports are constructed which are required to be dredged from time to time. The debris generated by dredging operations to maintain navigation channels is dumped into the open sea causing pollution. Some other problems related to shipping are - spilling of lubricants or liquid fuels, discharge of sewage and allied wastes etc. Plastic pollution due to merchant ships is taking a heavy toll of sea birds as once ingested plastic obstructs the digestive tract or causes ulcer in birds. Smaller plastic items are frequently mistaken as prey by turtles. Killing sea birds by sailors for food is also inevitable. Antifouling paints coated on the ship harm the coastal creatures even at some distance from the ship. Wastes from the periodic maintenance of ship by bottom scraping and sandblasting leave toxic material into the coastal environment.

Tourism

In India, promotion for tourism is given more and more emphasis which has resulted in too many large hotels along the beaches. These hotels and resorts dump their solid waste and sewage into the sea. Also, inception of such commercial establishments complies with an increase in population density around the coasts which also puts a burden of domestic waste on coastal areas. Aquatic life is harmed by the fishing for fun and increase in turbidity resulting from the stirring action of motor and speed boats. In Addition, leakages of lubricants and fuel from motor boats pollute the marine environment. Recreational activities such as snorkeling, scuba diving, spearfishing, shell collecting, coral walking by tourists and repeated anchoring of boats in reef areas cause significant damage to the corals. Hunting and egg-collecting cause declination of sea bird population. Many studies show that coastal nesting grounds of birds have been impaired by recreational activities.

Siltation

In India, five rivers on the east coast and two on the west coast discharge water which carry sediment transport from quarrying operations, land use changes etc. alongwith surface run off and joins the sea. A large quantity of sediments due to offshore drilling and construction both at sea and along the coast disturb the marine environment. Aquaculture activities and deforestation also lead to the siltation of ocean. Natural sources of siltation are coastal erosion and river debauching. Sediments decrease the penetration of sunlight, reducing the photosynthetic activity. Deposition of sediments on the marine flora and fauna is detrimental to aquatic life. Fertilizers used for agriculture and nutrient rich top soil are being washed into the sea by wind and heavy rains. This results in eutrophication combined with rising level of siltation.

Coastal pollution

The coastal waters have been in use for the disposal of pollutants as they provide a cheap method of waste disposal. The choice of selecting the sea for waste dumping is based on the belief that the coastal system has a great assimilation capacity. But in recent times, huge quantities of waste is being dumped which is much beyond the assimilation capacity of the coastal areas. In India, the prime source of the coastal pollution is from chemical industries. Pollution disturbs the coastal ecosystem, harms living resources, induces hazards to human health and a hindrance to the beneficial coastal activities. Various sources of the coastal pollution include domestic and industrial wastes, oil pollution, power plant outfall etc. as mentioned below.

Domestic Sewage

A high concentration of the coastal population (about 230 people per km) in India discharges domestic and other waste into the sea (UNEP, 1985). Untreated domestic sewage causes deposition of solids on the benthic flora and fauna, adding to the depletion of dissolved oxygen with ecological degradation. The discharge of municipal sewage is frequently accompanied by a decrease in salinity and corals are sensitive to changes in salinity. Observations show a steady declination in Chaetognath fauna population of Mumbai due to the sewage disposal (Nair *et al.,* 1991). Also dissolved toxic chemicals ingested by algae or zooplankton, entering the coastal food web leads to biomagnification. Pathogens present in the sewage produce biological imbalance to the coastal flora, fauna and human beings.

Industrial Wastes

Effluents from industries vary in quantity and quality. Even minor concentrations of Ammonia (from fertilizer waste), heavy metals and cyanide are potential hazards to fish. While alkyl benzene combined with paraffins cause biodegradation and mutation in the aquatic fauna, heavy metals like copper, cadmium, zinc, mercury are found to be accumulated in fish tissues known as bioaccumulation. The flyash clogs gills of the fish leading to its death. Observations show the accumulation of mercury in seafood from the Mumbai coast and higher metal pollution in the proximity of Lakshadweep and Andaman Islands. Scientists found that high DDT concentration in fish-eating birds produce thin shelled eggs. Such eggs break during incubation when the bird sits on it, affecting bird population largely. Industrial discharges into the rivers and other fresh water bodies also joins the sea affecting coastal resources.

Agriculture Wastes

Various types of agricultural wastes get run off during the monsoons. These are animal/cattle wastes, fertilizers and manures, pesticides, herbicides, insecticides, sediments from agriculture fields, crop residue and vegetable wastes. These wastes via food chain may affect human beings.

Oil pollution

About 3 million tonnes of oil products globally are lost into the coastal environment every year due to natural seeps, ocean dumping, shipping, mechanized fishing vessels, flushing and wreck of oil tankers. The toxicity of certain oils increases with storage and exposure to sunlight. Some of the oil dispersants are poisonous to the coastal life than oil itself. Residues of heavy oil sink to the bottom and degrade very slowly in the coastal sediments. Prevention of sunlight penetration by oil films lowers the primary productivity, which is the initial lead for entire coastal food web. Oil slicks inevitably cause sabotage to the spawning grounds, fishes, coral reefs, mangroves and sea birds. Oil is rapidly absorbed by suspended organic matter that is readily available to subsurface organism, and thus, make their entry into the coastal food chain. Tainting of oil on fishes cause bad taste or odour rendering them unpalatable. Consumption of these fishes are injurious to human health. Oil on the seasurface causes death of Cormorants, Pelicans and other diving birds. When birds plunge into the oil spilled water, their feathers that insulate them from the cold and buoy them up in the water are glued together, the birds can no longer fly. In the process of cleaning the oil coated feather, birds ingest a portion of it that often leads to their death.

Power plant outfall

Radioactivity is introduced into the coastal environment by the testing of nuclear weapons, accidents occurring in the reactor and power plants, and disposal of radioactive wastes. About 0.2% of radioactive particles find its path into the human body via coastal food web. Radiation from nuclear testing inevitably leads to widespread destruction of coral reefs and coastal animals. Increase in seasurface temperature due to the discharge of cooling water from power plants reduces the dissolved oxygen concentration. Thermal pollution can cause death to coastal organisms by extreme thermal shock, eutrophication, migration of zooplankton, obliterate metabolic rate and physiological functions.

Global Warming and Sea Level Rise

Many of the world's richest and mostly heavily populated areas and agricultural zones are in the low lying lands along the sea coasts. In total, it is estimated that about half of the human race lives in such areas. Many of the world's largest cities, including Mumbai, Calcutta, Shanghai, Bangkok, Jakarta, Tokyo, London, and New York are also in low-level coastal areas. Any rise in sea level increases the risk of flooding in these areas. Effects of sea level rise due to global warming when associated with fiercer storms and hurricanes may result in worst disasters. In Bangladesh, such storms that killed 250,000 people in 1970 and that which claimed a similar number of victims in 1991, were associated with cyclones. The projected rise in sea level indicate that one metre rise would impact about 360,000 km of coast line, render some island countries uninhabitable, displace tens of millions of people, threaten low lying urban areas, flood productive land and contaminate fresh water supplies.

Many third world countries rely heavily on the tourist trade, which would be affected as many beaches will be eroded due to sea level rise. As a rough average, a rise in sea level of 10 cm means a loss of about 10 m of beach. Intrusion of salt water into fresh water reserves can be another serious problem affecting agricultural, fishing and wild life habitats. About 10 cm rise will cause penetration of salt water about one km further inland.

Rising sea levels could out strip the rate at which the coral reefs are growing, thus reducing the degree of protection they provide to the coast. Mangroves also are sensitive to changes in sea level, as well as to the changes in water salinity and sedimentation rates, which are inevitable with rising sea levels. In testimony before a US senate committee in 1990, several scientists presented strong evidence that recent coral bleaching incidents around the world were directly related to abnormally high ocean temperatures caused by global warming. When the bleaching incidents were compared with global temperature charts for the past decade, a close correlation was found between years of high water temperatures and coral bleaching.

ISSUES OF MUMBAI'S URBAN ECOSYSTEM

Demographic and Economic Patterns

With a population of about 14 million (1.6% of India), Mumbai contributed around 4.5% of India's Gross Domestic Product in 1990-91. Of the total industrial product of India, contribution of Mumbai is around 9% (Tiwari, 1995). Urban sprawl of Mumbai is quite evident as from a built-up area of 234 km^2 in 1968 the city has grown to 575 km^2 in 1991 at a rate of around 5%, annually. The regulations governing land use have restricted vertical development and the growth is mostly horizontal. The population in Mumbai has grown at an annual rate of 1.87%. The distribution of population is more interesting. In 1991, 32% of the population has been in the island city, 42% in suburbs and 26% in extended suburbs. While the population in the island city has remained more or less constant since 1981 more due to policies than decline of the city. The suburbs have grown at 1.7% and extended suburbs at 6.05%. The employment of Mumbai

Metropolitan Region(MMR) which includes island city, suburbs and extended suburbs has grown at an annual compound rate of 1.35%.

The Process of Land Conversion In Mumbai

Analysis undertaken for the growth in population density of the MMR as per 1991 census and the estimates for 2001 and 2011 show the progressive 'Densification' of rural areas that take place as the urbanization proceeds (Table 3.1).

Table 3.1: Densification of Rural Areas (population in thousands)

Region	Area (sqkm)	1961	1971	1981	1991	2001	2011
City	49.9	2771	3070	3285	3174	3000	2825
Suburbs	141.0	1037	2167	3523	4168	4930	5910
Extended Suburbs	263.0	344	733	1436	2583	3500	4196

Source: BMR (1995)

It is important to have some understanding of the nature of this `densification' as a social process, and of the structure of the city which it yields. Mumbai is not simply a modern city marred by temporary slums but it is more or less permanently dualistic - the two components often being termed as 'formal' and 'informal' (Table 3.2). Perhaps more realistically, it is highly fragmented city composed of different communities pursuing different cultural goals with certain specific functional relationships.

Table 3.2: Employment in Formal and Informal Sectors

Year	Formal sector (Million)	Informal Sector (Million)
1971	1.13	1.07
1981	1.27	1.59
1991	1.18	2.25

Source: Census of India, 1991

A rough and ready typology of communities in Mumbai would include: affluent expatriates, professional elite, migrants from other states, domiciles (Maharashtrians) and rural-urban immigrants. Of course, the foreign and professional elite share lifestyles and, in spite of their very small numbers, determine almost wholly, by their lifestyles and their command of political and economic decision-making process, the general shape and structure of the city. Nevertheless, and in spite of an ongoing process of integration into the same classes of indigenous population, there are distinct affluent expatriates and professional elite lifestyles that express themselves in terms of specific locations and urban structures and facilities. These areas are well defined in terms of 'legal' land ownership title and relatively well served by publicly provided infrastructure.

The majority of the population has a very modest lifestyle but, because of their numbers, nevertheless, make a major impact on the city structure. Seen from the air, the city comprises a sea of concrete roof punctuated by the highways, high-rise offices and apartment blocks, universities and shopping centres of the rich. But internally these areas are differentiated. Starting from main streets lined with shops walks through back alleys reveal modest houses that become smaller and more dense as one proceeds away from the railway lines. Neatness gives way to less well kept paths with rotting refuge more prevalent. The creeks at Mahim, Thane and Vasai are sites of frequent refuge dumps. These are nevertheless the most densely populated areas; this is the habitat of new immigrants to the city.

Mumbai Metropolitan Region Development Authority (MMRDA) estimated the income distribution curve of different groups making up the city which indicates that in 1989, 62% of the households in MMR earn less than Rs.3230 per month. Around 25% of the households are below poverty line of Rs.1290 per month. It is clear that most of the population can contribute little in monetary terms towards the organization of formal urban services including water supply, sewerage and solid waste management. But income distribution in Mumbai is extremely skewed, with the average income lying between the 75th percentile. This implies a 'modern' city of between two and three million in a megacity of over 14 million. Modern city inhabitants can afford private services for water supply and refuge collection, with no questions asked with regard to where it all ends up.

The current rapid growth of Mumbai has encroached in the rural areas. The urban-rural population distribution in Mumbai is 92% urban and 8% rural. The countryside already has population densities not too far removed from those of the rich suburbs. The average land holding is less than 0.2 hectare. Impoverishment associated with ever-diminishing farm sizes encourages farmers to take up such opportunities in the urban economy as present themselves. In terms of use of their land this means mining it for building materials - bricks, sand, stone - if possible. Sooner or later it means selling out to speculators who, in the end, will expect to reap the profits of development. Vasai-Virar, Alibagh and Kalyan have already entered this speculative land market.

Generally, farmers are allowed to continue farming until developers decide to build. Massive formal sector projects are being realized in Vasai-Virar. These include a number of industrial estates of several hundred hectares each, cleared and waiting for factories to appear (common sight in the municipal corporations of Thane, Kalyan, Mira-Bhayandar, and Bhiwandi, and in New Mumbai, Pen, Khopoli and Alibagh). Large real estate development of row houses, including extensive recreational areas and a rash of golf courses. Pushed off the land, farmers' cluster into densifying rural settlements. They are increasingly joined by immigrants from other parts of country who in the past would have migrated into the city centre but who are now attracted directly by suburbs by the prospects of job in the large industrial plants as shown in Table 3.3.

Table 3.3: Spatial Distribution of Employment

Location	1971	1980	1990
Total (thousands)	1528	2199	2426
CBD	33.72%	32.15%	29.51%
Rest of Island	38.18%	31.47%	26.12%
Suburbs	28.1%	36.38%	44.38%

Source: BMR, 1994.

Environmental Issues in Mumbai

Not only large population but also its concentration in particular areas of the city has aggravated the environmental problems in the city. Some of these problems are as below.

Coastal Areas

Mumbai has 167 km long coastline that is highly indented with creeks, estuaries and bays. The coastline has undergone extensive changes during last several decades due to land reclamation. This has interfered with natural erosion process and is believed to have caused intense erosion that is witnessed in some parts of regions, such as Versova. Mumbai's wetlands have rich ecological diversity. Wetlands perform many useful functions such as flood control, shoreline stabilization, habitat to flora and fauna, retention of sediments, nutrients and toxicants. Urban sprawl in Mumbai is causing reclamation of wetlands, which are also converted into agricultural lands, salt panes, fisheries and solid waste dump ground.

Water Pollution

The levels of dissolved oxygen in water resources are much lower than the permissible standards. Most of the city's refuge is disposed off in its coastal water, which includes the coast line and six major creeks; Vasai, Manori, Malad, Mahim, Thane and Dharamtar. It receives 2181 mld/day of domestic waste with a total BOD load of 425 t/day, and 243 mld/day of industrial waste with BOD load of 24.3 t/day, bulk of which is untreated. This is increasingly deteriorating the quality of coastal water (Table 3.4). Although the region receives ample rainfall, water problems have escalated. Mumbai is facing acute water shortage (Table 3.5). Many areas in extended suburbs like Mira-Bhayandar, Vasai-Virar do not have adequate supply of water and have to resort to private borewells. Depleting aquifers coupled with reclamation of wetlands will surely cause salination of ground water, flooding and subsidence of these areas.

Table 3.4: Physio-chemical parameters in different coastal regions

Stations	Year	Salinity (ppt)	BOD(mg/l)	DO (mg/l)
Colaba	1986	36.3	0.49	5.61
(city)	1991	37.8	0.91	4.80
Thane Creek	1986	35.8	0.66	4.45
(Suburbs)	1991	36.4	3.2	3.9
Mahim Bay	1986	36.1	1.7	4.1
(City)	1991	36.9	3.1	3.7
Versova	1986	30.9	1.1	2.89
(Suburbs)	1991	32.9	2.8	1.9
Vasai (Suburbs)	1986	35.7	1.0	3.7
	1991	36.0	1.98	2.16

Table 3.5: Water Demand and Supply (1991)

Region	Demand (mld)	Supply(mld)
Greater Mumbai (including suburbs)	3026	2502
Western Region	102	35
North-East Region	883	823

Solid Waste

In spite of major efforts to organize the removal of refuge from the city and other urbanizing areas in the region, open spaces continue to accumulate piles of solid waste. Once predominantly organic in content, now solid waste, increasingly contains hazardous materials. MMR generates 4305 t/day of solid waste and 2485 t/day of debris. Generally refuge is collected from the common road-side collection points and is transported in trucks to the dumping grounds. Inadequate collection frequency, inaccessibility of certain areas, such as slums, throwing of refuge in open drains, gutters, and storm water channels. In addition, industries generate 508 t/day of solid waste. The solid waste dumping sites in Mumbai are located along creeks. At these sites, wastes are deposited below high tide levels without any soil cover. The total capacity of these dumping grounds is not more than 6000 t/day. Scavenging of solid waste at collection bins and dumping sites is carried out by waste pickers as many of them earn their livelihood from picking the recyclable waste. A recent study in Mumbai has found that about 14%, 58% and 25% of total earnings of a waste picker, on an average, comes from wastepaper, plastics and metallic waste. This results in various kinds of health hazard among waste pickers and creates unhealthy conditions (Sharma et al., 1997). Over utilization of dumping grounds and over polluted waste have caused irreparable damage to the coast by driving away fishes and damaging coastal biodiversity.

Air Pollution

Alarming levels of air pollutants are a serious threat to health and property in the city. Industrial and automobile emissions have reached a dangerous level and the city is considered as one of the most polluted cities in the world. Levels for some of the major pollutants are shown in Table 3.6. While sources of gaseous pollutants are well known, the suspended particulate matter (SPM) can be generated from enumerous sources. In an exhaustive study about the aerosols in Mumbai, it has been found that a large percentage of SPM is respirable causing various types of respiratory diseases (Sharma and Patil, 1992).

Table 3.6: Air Pollution Levels

Region	NO_x			SO_2			SPM		
	Max	Min	Avg	Max	Min	Avg	Max	Min	Avg
City	129	58	8	180	45	4	474	324	55
Western Suburbs	98	46	8	101	26	3	505	256	58
Eastern Suburbs	124	45	6	110	35	4	624	271	43

The Long Term Sustainability of Mumbai

'Sustainable Development' is development that meets the needs of the present without compromising the ability of the future generations to meet their own needs. The most intractable difficulty with the concept of sustainability is that there is no single criterion by which to measure it. Some of the questions and issues that might be relevant are given below.

- Is a city unsustainable when its surface water resources turn anaerobic and foul smelling and when piles of solid waste remain strewn around the streets? Many cities in the industrial world suffered such conditions for many decades, while continuing to grow, before steps were taken to improve matters.
- Is a city unsustainable when air pollution levels reach a point where on certain occasions the number of people dying as a consequence becomes significant? London suffered such conditions at various times in the history without yet reaching a point where its viability came into question.
- Is a city unsustainable when it runs into water shortages? People may learn to live using less water.
- Is a city unsustainable when it fails to receive sufficient food to feed its population? There have certainly been cases - as recently as in the China in 1960s-where this factor has been important contributor to urban depopulation.
- Is a city unsustainable when the export potential of its economic activities declines substantially? Such processes are in evidence in some Latin American cities today, leading to increasing impoverishment while urban growth continues.
- Is a city sustainable when it fails to receive the input of energy on which it has depended? This is an interesting question that has yet to be faced under contemporary conditions.
- Is a city unsustainable when global pollution leads to significant changes in climate? Such a situation is as yet unknown and the answer clearly depends on what changes in fact occur.

Probably, the most important sustainability criteria for a coastal city like Mumbai would be changes in climate and pressures of urban forces. The city is undergoing transition and this is an appropriate time to address revitalization of city.

Transition of Mumbai

Structural Transformation: Low value added jobs are moving out of city. New jobs being created are high value added and basically concentrated in industries like finance publishing and printing, software, electronics and similar sectors which are less polluting.

Coastal Zone: One study indicates that of the total coast line of Mumbai only 30% is accessible to public. Rest 70% are being used for some economic activity. Of late many environmental organizations have joined hands against pollution on coastal zones and beaches. The protest against Bandra-Kurla complex is one example. This trend might be important stimulator to protect ecosystems in Mumbai.

Poverty: Poverty has been the major cause of environmental degradation in coastal cities. Mumbai's 55 percent of population lives in squatter settlements which are characterized by poor water facility and drainage system. Mostly these settlements are along the coasts. Untreated discharge and unplanned settlements along the coasts have been detrimental to the eco-system. Recently, Government of Maharashtra has designed a scheme to resettle slum dwellers to formally constructed houses. This can be a significant Opportunity to protect coasts through the better settlement plans.

Comparative Economic Advantage: Traditionally Mumbai has always been an advantageous location for business to locate due to its infrastructure. However, after economic liberalization in 1991 this unique position has not remained unchallenged. Following 1991, the property prices zoomed by 200-300 % in four years due to speculative activities. This deterred new investments and better infrastructure of other cities have encouraged businesses to locate to alternate locations like Delhi, Bangalore, Pune, Hyderabad and Calcutta. These cities have upgraded their infrastructure and rival Mumbai. Due to loss of business prominence, Mumbai has seen downward slump in its property prices. Jobs are moving out of the city. Probably economic and infrastructure restructuring of the city might infuse new planning paradigm which is environmentally oriented.

Local Governance: Most municipalities in MMR are fiscal catastrophe. Bankrupt in terms of finances, these municipalities have practically nothing to improve the infrastructure. Municipalities have always prevented private participation in the provision of urban infrastructure. Liberalization of policies and bankruptcy have forced municipalities to invite private participation. Better city infrastructure and environs are advantageous for business as it would attract more investment. Some reforms like urban land ceiling act and zoning etc. have yet to take place. But this can be an opportunity for private sector to participate and influence policies, which can help city and its environs.

Political Systems Change: The present ruling party in the state is basically a regional party, for the first time, which is aware of city's problems. This can have both advantage and disadvantage. In past, Mumbai has contributed around half of the tax raised in this country. Only a small part has been transferred back to the state. With the regional party at the helm of affairs, one expects that more aware governance might influence a justifiable share to support city infrastructure and ecosystems. However, the danger is that the federal and State Governments have different ideologies, which may hamper the state's interest.

Social Change: During 1992-93 city saw worst ever religious riots in its history. The Cosmopolitan character of the city was torn apart. This, however demonstrated one fact that economic deprivation led to such a dire state of affairs. Aftermath has also been interesting and businesses like leather, carpet, trade etc. have moved out.

From Formal to Informal: As mentioned earlier, Mumbai is undergoing gentrification. Nineteen eighties saw textile industries going sick. Many manufacturing industries have also started moving out of the city. The city has changed its character to more of a trading and financial centre. Around 60% of workers are employed in informal sectors. An important consequence of this process of gentrification is that a huge chunk of land, which was occupied by these industries, is being made available for other environment friendly uses. Since industries used to be located near water bodies (textile industries in Parel and Dadar, for example) and disposed their wastes into them, the quality of these water bodies may improve in coming times.

Policy Resources in Hand: The transition of Mumbai has provided with following policy tools, which can be used to impact the deterioration of the city: Land regulations, Institutions, Capacity building/training, Fiscal (private versus public), Private industry/Multi-national companies concentration and Gentrification.

COASTAL ZONE MANAGEMENT

The activities and policies to conserve the resources of marine and coastal ecosystems to prevent further environmental degradation must be based on objective evaluations of such systems. Formulation of appropriate policies for efficient coastal zone management is itself a major area and requires in-depth studies of various aspects mentioned earlier. Some of the measures, which could be adopted, are mentioned as follows.

Measures for Local Problems

- Education and awareness- at all levels, schools, colleges, common man etc.
- Prevention and control of oil pollution - particularly in the Bay of Bengal and Arabian Sea. International water issues are crucial as pollution in one country may affect other countries.
- Declaring species rich areas as Biosphere Reserves or National parks.
- Control over exploitation of ocean resources.
- Improved fishing methods: Restriction to destructive fishing practices and over harvest, Prevention of mechanized crafts in species rich zones, Legislation to close the fishing operations in the nearshore waters during post monsoon period (peak period for abundant juveniles) for conservation of replenishment of stock for sustainable yield, Avoid using the nets with mesh size less than 50 mm to prevent the exploitation of juvenile stock, Setting up breeding centre for coastal fisheries etc.
- Development of artificial reefs for regeneration.
- Establishment of Research Centres to understand the various coastal ecosystem in a better way.
- Zoning for rare species, replenishment, harvest and tourism should be promoted.
- Restriction to deforestation, promoting afforestation, proper land use management to reduce siltation.
- Efforts to ensure effective treatment of industrial wastes to conform the effluent standards set by the Government.
- Progress in waste reduction, in reuse and recycling.
- Development of continuous monitoring stations along the coast.
- Absolute recognition and enforcement of the Law of the Sea.

Measures for Global Problems

Sea defense mechanisms such as dikes can be used to protect cities and coastal lands, as in the Netherlands, for example, over hundreds of years. But such measures require time, money and a reasonable degree of certainty about what is happening. The design and construction of hundreds of kilometers of dikes, drainage channels, pumping stations and the rest can only be under taken on a time scale of decades. The costs are also huge and far beyond the ability of most the third world countries to finance themselves. Immediate measures are required to reduce the green house gas emissions, which are the root cause of the global warming, and its impacts including sea level rise.

CONCLUSIONS

Studies on the coastal ecosystems and eco-management of the coastal cities are very important because these ecosystems are under threat from both land and coastal invasion by human settlements. Indian coastal ecosystems and coastal cities have deteriorated due to over-exploitation of resources, population influx and inadequate infrastructure. Coastal cities like Mumbai are undergoing social, economic and political transition. This is an appropriate time to rejuvenate these cities and protect them from further deterioration, otherwise, they will lose their comparative advantages to newer cities which have been more environmentally oriented. Another factor that coastal cities like Mumbai should worry about is the policy of reclaiming land. This is not only destroying the coastal biodiversity but also threatening the existence of city due to sea level rise and resulting submergence and flooding. A sincere and participatory efforts are required on every issue concerning coastal cities involving all stakeholders. An elaborate assessment of cost and benefit of each activity is required while developing infrastructure for the future.

REFERENCES

Ali D.M., Pandian, P.P., Somvanshi, V.S., John, M.E. and Reddy, K.S.N., (1991), Spear Lobster, Linuparus Somniosus, Berry & George, 1972 in the Andaman sea, Occasional papers of FSI, 13p.

Ansari, Z.A. and Abidi, S.A.H., (1989), Andaman sea - Its physical, chemical and biological characteristics, Management of aquatic Ecosystems, 21-32.

Ayyakkannu, K., (1990) , The Coleroon and adjoining seaboard (Southern India) a holistic study of benthic ecology, In: Environmental Research: Executive Summaries of completed projects, Vol.1, Ministry of Environment and Forests, New Delhi, 142p.

Bhattathiri, P.M.A., (1984), Primary production and some physical and chemical parameters of Laccadive and Andaman sea, Ph.D. Thesis, Mumbai University, 190p.

Blasco, F., (1975), The Mangroves of India, French Institute, Pondichery, Trav. Sec. Sci. Tech., Vol.14, 1-175.

Delfi Hydraulics and Tidal Waters Division, Rijkswaterstat (1993), Sea Level Rise A Global Vulnerability Assessment, The Netherlands.

Desai, B.N.,ParulekarA.H. and Wafar, M.V.M., (1992), Status of coral reefs of the Gulf of Kachchh and Andaman & Nicobar Islands, Nb, Gea, 52p.

FSI, (1992), Resources Information Series, Vol.3, No.3, 29p.

FSI, (1993), Resources Information Series, Vol.3, No.4, 25p.

Govt. of India, (1987), Mangroves in India: Status Report. Ministry of Environment and Forests, New Delhi.

Gross, M.T. and Kiemas, V., (1986), Use of AIS data to differentiate marsh vegetation, *Remote Sensing on Environment*, Vol.19, 97-103.

Foley Gerald [1991]:The Implications of Global Warming ,Chapter -3,Global Warming, pp37-50.

Jagtap, T.G., (1992), Coastal flora of Nicobar group of Islands in Andaman sea, *Indian Journal of Coastal Sciences*, Vol.21, 56-SB.

Jagtap, T.G. and Inamdar, S.N., (1991), Mapping seagrass meadows from the Lakshadweep Islands (India), using Aerial Photographs, *Journal of Indian Society of Remote Sensing*, Vol.19, 77-81.

Jyoti E.A. and Hair, V.R., (1990), Fishery potential of the Thane-Bassein creek system, *Journal of Indian Fisheries Association*, Vol.20, 7-10.

Kana Timothy W., Baca Bart J.and Williams Mark L.[1986]:Potential Impacts of Sea Level Rise On Wetlands Arond Charleston, South Carolina.

Krishnamoorthy, R., Bhattacharya, A., Hatarajan, T., (1993), Mangroves and coral reef mapping of South Andaman Islands through remote sensing, In: Sustainable management of coastal ecosystems, (Eds.) M.S.Swaminathan and R.Ramesh, Madras, 143-151.

MMR (1995): Draft Regional Plan for Mumbai Metropolitan Region, MMRDA, Mumbai.

MOEF (1987): Project Document of Ministry of Environment and Forests, Govt. of India, New Delhi., MPEDA, (1987), Statistics of coastal products exports 1987, Cochin.

Nair, V.R., Gajbhiye, S.N. and Desai, D.N., (1991), Effect of pollution on the distribution of chaetognaths in the nearshore waters of Mumbai, *Indian Journal of Coastal Sciences,* Vol.20, 43-48.

Nair, V.R., Nagasawa, S., Ramaiah, H. and Nemoto, T., (1992), Unusual thickening of collarette in Sagitta bedoti (Chaetgnatha) from the polluted environments of Mumbai coast, *Indian Journal of Coastal Sciences*, Vol.21, 296-299.

Nair, P.V.R. and Pillar, C.S.G., (1972), Primary productivity of some coral reefs in Indian seas. Proc. Symp Corals and Coral Reefs, Coastal Biological Association of India, 33-42.

Nayak, S., Pandeya, A., Gupta, M.C., Trivedi, C.R., Prasad, K.N.and Kadri S.A., (1988), Application of satellite data for monitoring degradation of tidal wetlands of Gulf of Kachchh, Western India.

Sharma V. K. and Patil R.S.(1992): Size Distribution of Atmospheric Aerosols and their Source Identification using Factor Analysis in Bombay, India, *Atmospheric Environment*, Vol. 26B, No.1, pp135-140.

Sharma Vinod K., Ghosh Upal and Haripriya G.S.(1997): Economic Analysis and Ecological Impacts of Fish and Shelfish Industry in India, *Encology*, Vol. 11, No. 10, PP 25-34.

Sharma Vinod K., Pieter van Beukering and Barnali Nag (1997): Environmental and Economic Policy Analysis of Wastepaper Trade and Recycling in India, *Resources, Conservation and Recycling,* Vol. 21 No. 1, pp. 55-70, 1997.

Tiwari Piyush (1995), "Bombay: A Snapshot History." Mimeograph.

Titus James G.[1987]:Greenhouse Effect ,Sea Level Rise,and Coastal Drainage Systems, *Journal of Water Resources Planning and Management* ,Vol.113,No.2 pp.216-227.

Titus James G.[1990]:Greenhouse effect, sea level and land use, *Land use policy*, pp138-153.

Titus James G.[1991]:Greenhouse effect and sea level rise:Potential loss of land and the cost of holding back the sea .pp1-45.

UNEP (1985): Environmental Problems of the coastal and coastal area of India: National Report, UNEP Regional Seas Reports and Studies No. 59.

Untawale, A.G., Wafar, S. and Jagtap, T.G., (1982), Application of remote sensing techniques to study the distribution of mangroves along the estuaries of Goa, In Wetlands: Ecology and Management, Proc. First International Wetland Conf., New Delhi, 51-67.

Varshney, P.K., Nair, V.R. and Abidi, S.A.H., (1983), Primary productivity in nearshore waters of Thai, Maharashtra coast, *Indian Journal of Coastal Sciences*, Vol.12, 27-30.

World Coast Conference (1993), Preparing to Meet the Coastal Challenges of the 21st Century: Conference Report, Hague, The Netherlands.

WWF, (1992): India's Wetlands, Mangroves and Coral Reefs, Report prepared by the World Wide Fund for Nature India for the Ministry of Environment and Forests, New Delhi, India.

COASTAL REGULATION ZONES AND DEVELOPMENT PROCESS

Vidyadhar Deshpande

INTRODUCTION

The process of urbanization has resulted in increasing the number of people living in urban areas. In case of Developed Countries this process was gradual and accelerated in 18th and 19th centuries. In case of Developing Countries it took place in last 5 to 6 decades. This process of urbanization, coupled with industrialization, transportation, tourism, recreation and many similar developmental activities, has resulted in affecting the vulnerable places. The most vulnerable places are the water bodies i.e. rivers, lakes, creeks, shore lines, coasts etc. To the large extent our coastal environmental ills are caused due to increasing demands and expansion of the activities. While these are important and inevitable for present generation we should not neglect the future generations. Healthy coastal life needs understanding and proper planning of environment, on and around the coast. Perhaps with these views only the Ministry of Environment and Forests, Government of India issued a Notification in the year 1991, under Environment Protection Act of 1986, declaring coastal stretches as Coastal Regulation Zones (CRZ) and regulating activities in the CRZ.

COASTAL REGULATION ZONES

For regulating coastal activities, the coastal stretches within 500 meters of High Tide Line (H.T.L). on the land ward side are classified into the following different categories of coastal regulation zones (CRZ).

Category - I (CRZ - I)

i) The areas that are ecologically sensitive and important, Such as National Parks, marine Parks, Sanctuaries, reserve Forests, mangroves coral reefs, areas close to breeding and spawning grounds of fish and other marine life, Historical heritage areas, and areas likely to be inundated due to rise in sea level due to global warming.

ii) Area between the Low Tide Line and the High Tide Line.

Category - II (CRZ - II)

The areas that have already been developed upto and close to the shoreline. For this purpose developed area is referred to as that area within the municipal limits or in other legally designated urban areas which is already substantially built up and which has been provided with drainage and approach roads and other infrastructure facilities, such as water supply and sewerage mains.

Some development on land ward side of existing roads and structures and proposed roads shown on the Coastal Zone Management Plan (CZMP) are permissible in this zone.

Category - III (CRZ - III)

The areas that are relatively undisturbed and those which do not belong to either Category-I or II. These will include coastal zone in the rural areas (developed and undeveloped) and also areas within Municipal limits or in other legally designated urban areas, which are not substantially built up.

Category - IV (CRZ - IV)

Coastal stretches in the Andaman and Nicobar, Lakshadweep and Small islands except those designated as CRZ-I, CRZ-II and CRZ-III .

The government regulations prohibit certain activities and also list the permissible activities within the CRZ.

Prohibited Activities

a. Setting up of new industry and it's expansion.

b. Manufacture, handling, storage or disposal of hazardous substances.

c. Setting up and expansion of fish processing units including ware housing.

d. Setting up and expansion of units for disposal of waste and effluents except facilities required for disposal of treated effluent.

e. Discharge of untreated wastes and effluents from industries, cities or towns and other human settlements.

f. Dumping of city or Town waste for the purposes of land filling.

g. Dumping of ash or any wastes from Thermal Power Stations.

h. Land reclamation.

i. Mining of sand and rocks.

j. Construction activities in ecologically sensitive area etc.

Permissible Activities

a. Those activities, which need water front and foreshore.

b. Certain activities will need Environmental Clearance from Ministry of Environment and Forests, Govt. of India, Listed below.

   i. Construction activities related to Defense requirements needing foreshore.

   ii. Operational constructions for parts and harbours and light houses requiring water frontage, jetties, wharves, quays, slipways etc.

   iii. Thermal Power Plants.

   iv. All other activities with investment exceeding rupees fifty million.

Implementation of CRZ Notification and its Implications

In Maharashtra, with its coastal line of 720 Kms. and 54 river creeks, significant stretches in land are badly hit by the CRZ Notification. It is posing several problems before the planners, decision makers and policy makers on one side and investors and developers on the other. Maharashtra is one of the leading states in the country, so far as industrialisation, urbanization, road and rail network are concerned. Metro City Mumbai, which is on the western coast, is considered as commercial capital of India and is destined to be one of the Megacities of the future world.

The CRZ Notification as mentioned earlier has put too many restrictions on the development along the Coast. Various issues of CRZ involved in the process of planning and development are as follows:

1) Definition of High Tide

High Tide line (HTL) in the CRZ Notification is defined as line upto which highest high tide reaches at spring tides. Such high tide occurs once in 13 years and it is very difficult to mark this line on the plan and guide the development according to CRZ Notification. The Chief Hydrographer to Government of India is the best authority, who can demarcate the high tide line, and who has got the proper expertise to demarcate such HTL.

2) Dealing with the Creeks

In case of creeks, CRZ Notification and subsequent Supreme Court Judgement have stipulated that if the width of the creek is upto 350 meters, the CRZ will be 100 meters from the creek and if the width exceeds 350 meters, it will be 150 meters from the creek. Now, from where to measure the distance of 100 meters or 150 meters is a point of debate. Experts in the field of hydrography say that it is possible to demarcate edge of the Creek and in that case this distance can be measured from the edge. Others say that in case of creeks, even the area up to the water spreads will form coastal regulation zone. If the second version is accepted, then naturally vast area in which water spreads, is excluded from the process of development. This brings lot of pressure on the existing land, which can be made available for development, particularly in urban areas like Mumbai, and other cities along the Coast.

Another point in respect of creeks is upto what distance inside the creek the CRZ Notification apply. British experts have opined that distance inside Creek, where tidal effect of the wave comes to and end, should be considered as the distance inside creek, for determining Coastal Zone.

3) Conditions Regarding Mangroves

It is stipulated that in case of mangroves with an area of 1000 m^2 or more, would be classified as CRZ with a buffer zone of at least 50 meters. Mangrove is a tropical tree growing along the coast and requires saline water for its growth. Experts in the field say that mangroves are very important along the coast for breaking of tides and it is valuable resource having several direct uses, and hence it is needless to say that protection of mangroves is very important. Nevertheless, distance of 50 meters is quite large and it can be reduced to 15 m. particularly in urban areas where land values are high.

4) Limit on the Project Cost

CRZ Notification specifies that clearance from the Ministry of Environment and Forests should be taken, if the cost of the project planned within CRZ is more than Rs. 50 Million. It is not stated whether the cost of the land should also be included in the cost of the project. If this is so, it is natural, that in case of Mumbai or many other cities, which are situated along the coast, where land values are sky rocketing the cost of even a small project is expected to exceed the limit of Rs. 50 millions. Hence, it is necessary that this limit should be different in different areas, depending upon the land values or else the value of the land should be excluded from the cost of the Project.

5) Slum Redevelopment Schemes of Government of Maharashtra

In Mumbai there are many places where slums have come up in Coastal areas and in quite a few areas right upto the water edge of the creek or main coast line. Solid and liquid wastes as well as human excreta are generously disposed of in coastal area or into the inter-tidal areas, which falls in CRZ-I category, or in the nearest creek. Years of long experience have made it clear that it is preferable to reconstruct buildings at the same site rather than shifting these hundreds or thousands of slum families elsewhere. However, due to CRZ Notification it will not be possible to rehabilitate these people at the same place. It is, therefore, necessary that planners should take this issue to the Central Government to allow slum redevelopment schemes of the Government as an exception to the CRZ Notification.

6) Proposed Roads

As per CRZ Notification, in CRZ-II it is stated that buildings shall be permitted only on the landward side of the existing and proposed roads, existing authorised structures shall be subjected to the existing local Town and Country Planning Regulations. It is mandatory for the local Planning Authorities to prepare development plans of the towns in their jurisdiction as per the provisions of the Maharashtra Regional and Town Planning (MRTP) Act, 1966. Many towns including Mumbai have prepared development plans accordingly and the development is carried out as it is contemplated in the development plan. In case of Mumbai, the development plan was prepared and sanctioned in the year 1966 and was subsequently revised in the year 1984 and was sanctioned after following procedure prescribed in the MRTP Act. It would, therefore, be proper to take into account the proposals of the development plan, which was sanctioned much prior to the date of CRZ notification. In that context, allowing the development on the landward side of the proposed roads in the development plan will not violate CRZ notification.

7) Special Status for Mumbai and other Coastal Cities

As stated hereinabove, Mumbai is destined to be one of the Megacities of the future world and is under tremendous pressure of development. It is, therefore, needless to say that special treatment will have to be given to Mumbai, considering its problems and its special characteristics as commercial capital of India. There may be similar cities like Mumbai along the Coast needing such special treatment. The Ministry of Environment and Forests should, therefore, think appointing special commission for considering special cities like Mumbai and other important cities along the coast.

8) Permitting Recreational and Entertainment Activities in CRZ-III

It is stipulated while sanctioning the Coastal Zone Management Plan of Maharashtra that all Parks, Play Grounds, Recreation Parks, Green Zones and other similar areas falling within the CRZ-II areas are categorised as CRZ-III. Not permitting such activities goes against the proposals of the development plan.

9) Projects of Public Importance

There are certain projects, which are of public importance in Mumbai such as Bandra-Worli Express way, Nhava-Sheva Seweri Sea Link. These projects are of vital importance so far as the traffic needs of Mumbai and surrounding area is concerned. Such Projects should be given utmost priority and need clearances from the concerned authorities as early as possible.

CONCLUSIONS

While carrying out development, which is important and vital for the present generation, we should not forget the interest of the future generations. However, at the same time we should not get too bogged down by future needs putting complete halt to the process of development. Development plans should be such that risk of the future generations, due to development undertaken, is minimized. The plans should give us developed coasts with less disturbance to the ecological balance rather than to have undeveloped coasts, rendering them free ground for encroachers. Coastarica is the best example so far as Coastal Management is concerned. Indonesia has also developed its coasts in best way. Development should be carried out in such a way, that the negative impacts on environment are avoided. We have to think judiciously and take decision so that present generation satisfies its need and at the same time posterity is not affected adversely.

URBANIZATION AND INDUSTRIALIZATION AND THEIR IMPACT ON FISHERY RESOURCES IN INDIA

George J. P., O. Sudhakar and K.Kishore

ABSTRACT

Presently about fifty percent of the world population lives in the coastal areas. It is estimated that by the turn of the century over 3025 million people would be dwelling in the coastal regions of the world. There will be an increase of a billion in urban population by the year 2020 and this increase mostly will be anticipated in the third world. It is projected that Mumbai will be the second largest city of the world by the year 2000 and there will be three Indian cities (two on the coast) with a population of 10 million and over. Even before the dawn of the industrial revolution the inland and marine waters were always considered as dumping grounds. The growth rate of marine fisheries in India in the last few decades was not impressive. This declining growth rate has been presumed due to over exploitation of resources, however increasing pollution in the coastal waters can be another reason. The present article portrays the ecological consequences of urbanization and industrialization in the coastal areas of the country in relation to fishery resources.

INTRODUCTION

Due to the anthropogenic activities 6.5 million tonnes of litter find its way in to the oceans every year. Several million tonnes of top soil with a fertilizer value of about Rupees one billion is discharged into the sea due to the erosion. Over 150,000 tonnes of non-biodegradable plastic is dumped into the sea. Besides maritime activities, the growth of tourism industry has been responsible for several million tonnes of consumer plastic wastes entering the oceans. Eighty percent of the pollutants is untreated domestic sewage and remaining twenty percent is from industrial effluents. Pathogens in the sewage accumulate in fish and shell fish. Sewage disturbs benthic community and may even eliminate them. Excess of nutrients leads to extreme eutrophication and the occurrence of algal blooms. Many of the algal blooms are associated with production of algal toxins which are detrimental to the aquatic organisms. Industrial wastes are generally loaded with toxic substances such as acids, alkalies and ions of heavy metals. The polychlorinated biphenyls (PCBs) used extensively in plastic and rubber industries are especially known for their toxicity and non-biodegradable qualities. The wastes of highly diversified petrochemical industry are either highly carcinogenic or they spoil the flavour of the fishery products. Bioaccumulation of heavy metals and radioactive wastes may lead to complex abnormalities of the organism.

India has a long coast line and 2.02 million sq. km of Exclusive Economic Zone (EEZ) with replenishable marine living resource potential of about 3.9 million tonnes of fish. However only 2.2 million tonnes i.e. only 69% of the potential resources is being exploited. Apart from this, the annual growth rates of the marine fishery are sharply declining. In 1991-92 the growth rate was 6.39%, in 1992-93 it was 5.27% while in 1993-94 the growth rate was only 2.83%. In 1994-95 it drastically declined to 1.62% and an all-time low of 0.55% was observed in 1995-96 (GOI, 1996). The steep declining trend in the growth rate of marine fishery is mainly attributed to over exploitation; general marine environmental degradation especially in the coastal areas is also contributed to certain extent in the dwindling of marine catches in the recent past.

In the coastal areas of maritime states of the country there are a number of towns/cities, which are densely populated and industrialized. Apart from the three major cities of Mumbai, Calcutta and Chennai many other cities like Cochin, Visakhapatnam and Veraval etc. with dense population contribute large volumes of sewage and other domestic wastes. About 26% of the population lives in the urban areas of India. Undoubtedly, cities remain centre for economic growth and social development, but it is also true that in developing countries about 600 million of 1.7 billion urban residents do not have the basic needs for shelter, water and health and their environment is a dumping ground of wastes. Rapid urbanization and industrialization within coastal areas lead to the formation of slums which results further degradation of the coastal environment. The cottage industries associated with fisheries and other activities add further dimensions to the coastal pollution. In this paper the emphasis will be on the coastal pollution caused by urbanization and industrialization and their impact on fisheries resource of the country .

URBANIZATION, INDUSTRIALIZATION AND POLLUTION

Urban population increases rapidly as industry develops resulting in environmental pollution from trade and domestic effluents. Coastal pollution can seriously harm the aquatic life in shallow areas of the seas besides creating ugly and malodorous conditions which impair recreation facilities and spoil civic areas of tourist attraction.

Sources and types of pollutants from cities and industries are as follows:

a. Domestic Sewage
b. Marine Trade
c. Electrical power generation
d. Insecticides and Pesticides
e. Radio active Isotopes

Effect of Domestic sewage on fisheries

The anthropogenic influences on the coastal environment include all significant, observable changes in traditional fishing grounds that have occurred because of the excess of presence of our species *Homo Sapiens.* Domestic wastes usually discharge in untreated conditions into the coastal waters due to the lack of sufficient treatment facilities in most of the coastal cities and towns. It has been reported that only primary treatment facilities are available in cities and towns where population is more than 100,000 and the capacity of the plants is not adequate for the treatment of total waste generated in the city(Subramanian, 1993). Due to such partial treatment the chemical characteristics of the waste water retain almost their original features and cause damage to the water quality and the aquatic organism. In a coastal city, since the immediate discharge point invariably is the coastal environment, the fish fauna in the shallow areas which mostly comes under traditional fishing zone is affected adversely and some times large mortality occurs.

Most of domestic waste comprises phosphate based detergents. Phosphate, nitrate and silicate nutrient load results in explosion of algal growth. Some of the algae such as diatoms have given trouble in nutrient rich areas because they clump together on the gills of fish, preventing effective respiration. Some kinds of algae, eaten by shellfish produce toxins, which are harmless to shellfish but cause fatal paralytic poisoning of the people who eat them. (Edmondson, 1995). In addition, the beaches are considerably misused which not only spoil the aesthetic beauty, but also result in soil erosion, coastal flooding, increase in turbidity of water etc. Due to human activity 6.5 million tonnes of litter find its way into the oceans each year. Also, over 150,000 tonnes of non-biodegradable plastic is dumped into the sea annually (D'Souja, 1996).

The anthropogenically dissolved and suspended organic matter from the coastal areas can cause a steep rise in BOD resulting in an anaerobic condition and subsequent mortality of plankton, nekton and fish. The available oxygen is consumed for decomposition of organic matter causing oxygen depletion. With putrefaction of organic matter by the anaerobic bacteria Hydrogen Sulfide gas is released. It is toxic to the fish and other marine living organisms. Suspended solids increases the turbidity of the water which in turn cause excess of mucus secretion in fish and very high turbidity will lead to the clogging of the gills which cause asphyxiation even though there may be adequate dissolved oxygen in the water. Domestic sewage also lead to the siltation which destroys the feeding and breeding grounds of the fish and shell fish fauna of the area. Silt suspended in the water reduces the light penetration, which in turn affects the photosynthetic activity and reduces the productivity of the water body in terms of fish-food organisms resulting in dwindling the fish population. Anaerobic conditions created by the decaying organic matter will harbour anaerobic bacteria, which is harmful to the fish health.

Excess of nutrients like nitrates in domestic sewage have adverse effects on both human beings and living marine organisms. In human beings it causes Methaeglobaemia, N-nitrosamine is formed by simultaneous ingestion of nitrate and secondary amines and it is carcinogenic to marine living beings (Malve,1996). Non-degradable plastic materials from domestic source mask the benthic area ultimately leading to elimination of benthic communities which is an important component in the fish food reserve in aquatic ecosystem.

Marine Trade

Trade through the sea routes leads to the building up of several ports and movements of large cargo ships and oil tankers. The debris dredged to built and maintain navigation channels is being dumped into the sea. This debris not only disturbs the feeding and spawning grounds of the fish but also increases the turbidity of water that affects the density of the phytoplankton population and its exposure to light. The problem is further complicated and it affects the basic food chain in the fish food cycle and reduce the photosynthetic oxygen production. The settling of degradable and non-degradable materials can also destruct the coastal fisheries environment. Oil spilled accidentally and discharged after usage is one of the major problems arising out of the marine trade. The indiscriminate spilling of the used lubricant oil and its formation of a thin film over the gills of fish leads to suffocation and mass mortality. Garbage disposal into the sea by the ships and also from coastal inhabitants interfere in the fish food-cycle and ultimately affects the coastal aquatic ecosystems.

Power Generation

Rapid urbanization and industrialization has created a need for higher electricity generation. This need is being met through setting up of several thermal and nuclear power plants. The discharge of unutilised heat constitutes 70 % of the total heat produced in a thermal power station. In a nuclear power plant it is slightly higher. The discharge of the unutilised heat from the power plants into the cooling water source and the subsequent consequences in the ecosystem such as - stratification, potential process of uptake and degradation of organic matter are causing serious concern. The discharge of heated effluents from the power plants at 8 to 10 degrees higher than the coolant intake waters causes thermal stratification and directly inhibit the photosynthetic activity which in turn degrade the biogenic capacity of the coastal waters and fish production. Abrupt changes in the thermal regime of coastal waters can affect the metabolic as well as other life activities of aquatic fauna.

Effects of Thermal pollution on Fishes: Vertical distribution of marine organisms may be affected by temperature rise in three ways:

a. Exclusion of certain species from unsuitable levels;
b. Migration to suitable levels; and
c. Movements influenced by the change in viscosity and density.

Marine organisms are poikilothermic. i.e., their body temperatures are close to that of surrounding waters and varies accordingly. If any alteration of temperature takes place they will migrate to the areas where favorable conditions prevail. The thermal pollution if above 40°C it would kill most aquatic life. Cottage industries flourished in the coastal belt and discharges consisting various types of solid and liquid wastes discharged into the water, which is certainly affecting the normal life activities of marine organisms. Fishes show a marked rise in basal rate of metabolism with temperature to the lethal point. The uptake of food, respiratory rate and swimming speed increase with the rise of temperature.

The rising temperature triggers deposition of eggs by female fishes. The triggering is particularly dramatic in estuarine shell fish, which spawn within four hours of the water temperature reaching critical level. The Atlantic salmon eggs hatch in 114 days in winter (2°C) and 90 days at 7°C. Harring eggs hatch in 47 days at 0°C and in 8 days at 14.5°C. These are critical temperatures for reproduction. High temperatures may induce increase in activity, which exhausts the organism and shortens life span. Table 5.1 shows the thermal death point for some fish species.

Insecticides and Pesticides

Pesticides are biological toxicants and are known to bring diverse effect on aquatic organisms especially fish. The organophosphorus insecticides which enter into the water may decrease the level of carbohydrates, glycogen and total protein(Das,1989). Pesticide consumption in India has shown a fast increase during the last few decades. From 432 tonnes in 1954 it rose to 8,600 tonnes in 1961, to 24,820 tonnes in 1971 and nearly 80,000 tonnes in 1981. Pesticides leave residues which are responsible for disturbing the ecological balance. Damage to environment by pesticide residues results in two types of interactions: i) with biotic component, ii) with abiotic component.

Pesticides may pollute water through run-off from treated areas; air drift during the aerial application, industrial effluents etc. Pesticides in water cause the death of several independent aquatic forms of life and also a source of bio-magnification of persistent pesticides. Several reports of large fish kills due to pesticides, particularly endrin and DDT, are on the records. Toxicity studies in fish reported from India revealed greater accumulation of organo chlorines in tissues. It has been reported that the plankton samples from Arabian sea between Mumbai and Goa contain high levels of the DDT (Attri, 1981).

Table 5.1: Optimum Temperature and Thermal Death Point for Species

Fish Species	Acclimation Temperature °C	Thermal Death Point °C
Brook stickback	25-26	30.6
Brown Trout	26	26
Brown Trout fry	20	23
Carp	20	31.8
Cat fish	15	31.8
Chinook salmon fry	15	25
Chim salmon fry	15	23.1
Common Shiner	15	30.3
Common sucker	15	29.3
Creek chub	15	29.3
Pathead minnow	10	28.2
Golden shiner	15	30.5
Gold fish	30	34
Guppy	30	34
Perch	--	23.2
Pink salmon fry	10	22.5
Roach	20	29.5

(after Jones, 1964)

Radioactive Isotope Bioaccumulation

Radioactive metals and iodine which derive as a by product of nuclear detonation may result in bio-accumulation in the successive links of the food chain in aquatic ecosystem which ultimately find their way in to human beings through fish and sea food consumption. This may damage the normal metabolism and growth (Das, 1989). Radioactive pollutants cause genetic mutation in fish and marine living organisms.

IMPACT OF POLLUTION ON FISHERIES

As mentioned above, the marine ecosystem is under environmental stress due to several man-made activities. As a result of this stress, many of the identified fishing species are declining both in numbers and quantity. The Whale shark, *Rynodon typus* of Gujarat, the cat fishes of the genus *Techysurus* of Karnataka and white fish *Lactraius lactarius* along south west coast of India have been reduced in abundance. The once existing fisheries for *Polynemus indicus, P.heptadactylus, Pomadasys hasta, Utolithiodes bruneus, Protonibea diacanthus, Congresox talebanoides, Muraenesox cinereus* all off the Gujarat - Maharashtra coast and *Platycephalus maculipinna* along the south east and south west coast have become non existent at present (James, 1972).

CONCLUSION

India comprises only 2.6% of the total land mass of the globe which is occupied by 16% of the total world population out of which a large percentage is congregated in the coastal areas. Coastal pollution in India, due to the anthropogenic activities, is increasing at an alarming rate. The general productivity of the coastal areas came down due to the discharge of various types of pollutants which has to certain extent affected the fish population in the shallow areas of Indian seas. In view of this, and as we are aware of the pollutants and technologies have been developed for the pollution abatement, coastal environment monitoring and laws against pollution have to be enforced for the conservation of marine living resources of EEZ. It is necessary to measure the cost of prevention and control of coastal pollution against its effects. The effects should include not only human health but also the damage to flora and fauna in the coastal waters and perhaps the unquantifiable contribution of pollution to the deterioration of the quality of life of vast coastal population. Over and above, an awareness has to be created among the coastal population and educate them about the evil effects of the pollution causing disease, genetic disorders etc. before the coastal environment further deteriorate and reach a point of no return. This will ensure the availability of a hygienic, healthy coastal zones for the inhabitants in the next millennium.

REFERENCES

Attri, B.S., (1981), Status of Environment Pollution by Pesticides in India, In *Proc. Nat. Sympos. Evalu. Environ.* (Spl. Vol. GEOBIOS), p 33-40.

Das, S. M., (1989), *Handbook Limnology and Pollution,* South Asian Publishers, New Delhi, pp 117-127.

D'souja, Joe., (1996) ,By Poisoning the seas we are killing ourselves. In *Voices for the Ocean* : a report to the Independent World Commission on the Oceans,Ed. R.Rajgopalan.

Edmendoson, W.T.,(1994), Eutrophication In : *Encyclopedia of Evironmental Biology,* Volume 2 (F.N.). Academic press, London,p 697-703.

Gesamp. (1990), Human Activities affecting the sea In *The state of Marine Environment,* Blackwell Scientific Publications,pp 9-37.

GOI, (1996): *Handbook of Fishery Statistics*, Ministry of Agriculture and cooperation(Govt. of India), New Delhi.

James, P.S.B.R., (1972), Endangered, Vulnerable and rare marine fishes and animals, *Abstract 35. National Seminar on Endangered fishes of India,* NBFGR, Allahabad & NATCON Muzaffarnagar, April 1972 , pp. 47.

Jones, J.R.E,(1964) ,*Fish and river pollution,* Butterworth and co.(Pub) Ltd, pp. 153-168

Malve,S.P., & (1996), Nitrate and Environmental pollution, S.S. Dhage,*In Every Mans Science.*,Vol. XXXI. No. 5 pp. 158-163.

Subramanian,B.R. &(1993) *Marine Pollution and Coastal Zone,* S.A.H. Abidi, Management *In Environmental Impact on Aquatic and Terrestrial Habitats*. Society of Biosciences, 25/4 Ram Bagh Road, Muzaffarnagar. pp. 251.

6

CONSERVATION OF COASTAL BIODIVERSITY OF THE ISLAND CITY OF MUMBAI

Sanjay Deshmukh , Madhavi Inamdar, Nishikant Dighe,
Arun B. Inamdar and S.B. Chaphekar

INTRODUCTION

It is now widely recognised that coastal biodiversity is under severe threat due to urban expansion, industrialisation, pollution, tourism and other associated human activities. Tropical tidal forests, i.e., mangroves and other associated coastal ecosystems, such as seagrass and coral reef ecosystems are particularly becoming vulnerable due to expansion of aquaculture, chemical and petroleum industries, tourism and human habitation. The non-legally binding authoritative statement of principles for a global consensus on the management, conservation and sustainable development of all types of forests adopted at Rio De Janeiro in June 1992, has laid great stress on the sustainable management of forests and on providing for their multiple and complementary functions and use. The Rio Forestry principles also call for a holistic and balanced consideration of forestry issues. It is in this context that development of strategies for need based research dealing with the conservation, eco-redevelopment, and sustainable utilisation of coastal ecosystems, particularly mangrove ecosystems in India, assumes urgency and importance.

Mumbai can be considered as a region where mangrove ecosystems have been under serious threat due to both, biotic and abiotic factors. However, in many places in and around Mumbai, habitat conditions are still capable of supporting mangroves. Barring the odd exception of the work on ecology of Ulhas river (Aswani Kumar and Chaphekar, 1987) as well as ecology and eco-development of the mangroves of Mumbai by Deshmukh (1990), there is practically no record on the similar work along the Mumbai coast. Moreover, there has been no updating on the information on "status of coastal biodiversity in and around Mumbai."

METHODOLOGY

The preparation of the report to know state of the art of the coastal habitats of Mumbai and its surrounding region using remote sensing data and field surveys was undertaken during September 1996 and April 1997. Remotely sensed data of Mumbai and its surrounding coastal region was obtained from the National Remote Sensing Agency (NRSA), Hyderabad. Maps of this region (to the scale of 1:25,000) were prepared at the CSRE, IIT (Mumbai) by interpretation of satellite imageries (IRS 1C satellite FCCs , as well as SPOT data) and ground truth data.

Mangrove vegetation was studied for floral elements and their distribution along the coastal area of Mumbai, such as Ghodbunder (Thane district), Thane, Mulund, Vikhroli, Chembur, Trombay, Sewri, Colaba, Elephanta island, Mahim, Bandra, Mudh, Malad, Manori and Vasai. Other research components of the work are highlighted below:

1. To prepare an up-to-date inventory of mangrove and other coastal habitats of Mumbai (mainland and adjacent islands) based on field surveys.
2. To identify coastal wetland and/or landform categories of Mumbai island and to mark them on maps (1:25,000 scale) based on interpretation of the remote sensing data.
3. Integrate remote sensing based inputs with environment related data to identify suitable areas for *in situ* conservation of coastal biodiversity.

OBSERVATIONS

The city of Mumbai is an elongated trapezoidal area situated on a group of seven islands between 18°55' N to 19°20' N latitude and 72°45' E and 73°00' E longitude. This mainland has at its northern end the Ulhas river and on the west the Arabian sea. The coastline is indented by many bays, creeks and tidal inlets. The coastline on the west has three major creeks: Manori, Malad and Mahim, and three bays, viz., Mahim, Worli and Backbay. Extending from the head of Mumbai harbour bay (located at the south-eastern end of the city) to the Ulhas river in the north, is the Thane creek. The total expanse of the Greater Bombay city is 481.74 km^2.

Observations based on remote sensing studies are highlighted with the help of coastal wetland maps. Various wetland types such as estuary, lagoon, creek, bay, mudflats, sandy beaches, rocky shores, mangrove vegetation (dense and sparse), salt marsh and grass/ scrub vegetation, etc., are highlighted on the maps with different legends. Observations on the coastal wetlands based on field surveys are highlighted as follows:

MANGROVE WETLANDS OF MUMBAI

Ghodbunder: On the eastern edge, there are shorter creeks and on the protected side of the Mumbai island, the creeks extend gently into Thane creek. The mangroves of this coastline can be seen up to Ghodbunder at their northern side and upto Colaba, at the southernmost limit. As seen from the Table 6.1, most of the mangrove species present in Mumbai can be seen in this region. Being a relatively less disturbed area, a zonal pattern can be observed; there is a profusion though in patches, of *Rhizophora mucronata* towards the seaward fringe which extends up to the Gaimukh bundar and occurs in patches till Kasheli bridge. Further along the coast, *Avicennia marina* becomes prominent.

Thane: As one proceeds from Kasheli bridge to Thane creek, the abundance of *Sonneratia apetala* becomes evident. The mixed stands of *Avicennia marina, Sesuvium portulacastrum, Salvadora persica* and *Aeluropus lagopoides* continue till the Railway bridge near Thane where there are small islands on which *Avicennia, Acanthus* and *Sesuvium* are found.

Mulund to Vikhroli: The marsh tract from Thane extends towards the south along Thane creek where the small channels of the creek in this region form a network among the continuous stretches of mangrove vegetation and comprises a diversity of species. The vegetation is mainly represented by *Avicennia marina* and is accompanied by other species such as *Excoecaria agallocha, Acanthus ilicifolius* and *Sesuvium portulacastrum* while *Bruguiera cylindrica* appears to be common in the swampy region at Vikhroli. This region also comprises a few stands of *Sonneratia apetala.*

Table 6.1: Distribution of Mangroves and Their Associates at Various Locations in Mumbai.

Name of species	Location
Acanthus ilicifolius	1,2,3,4,10,12,14,15,16
Aegiceras corniculatum	1,4,12
Aeluropus lagopoides	1,2,3,4,6,14,16
Avicennia marina	1,2,3,4,5,6,7,8,9,10,11,12,13,14,15,16
Avicennia officinalis	1,2,5
Bruguiera cylindrica	4,12
Bruguiera gymnorrhiza	1
Ceriops tagal	1,4,12,16
Clerodendrum inerme	1,2,3,4,10,12,13,14,15,16
Cyperus rotundus	1,4
Derris heterophylla	2,3,4,10,15
Excoecaria agallocha	1,2,3,4,16
Kandellia candel	13
Rhizophora mucronata	1,6,8,15,16
Salicornia bracheata	1,7,13
Salvadora persica	1,2,3,4,9,10,16
Sesuvium portulacastrum	1,2,3,4,10,13,14
Sonneratia alba	8,10,15
Sonneratia apetala	1,2,3,4
Suaeda fruticosa	2,6,7,12,13,14,16

1: Ghodbunder; 2: Thane; 3: Mulund; 4: Vikhroli; 5: Chembur; 6: Trombay; 7: Sewri; 8: Colaba; 9: Butcher island; 10: Elephanta island; 11: Backbay; 12: Mahim; 13: Bandra; 14: Madh; 15: Malad; 16: Manori.

A few thousand saplings of *Rhizophora mucronata* were introduced in this region, thanks to the efforts at the Soonabai Pirojsha Godrej Foundation, Mumbai which, in mid-eighties conducted large scale plantation of mangroves in its protected area of Pirojshanagar, a 4000 acre coastal area under mangrove in Vikhroli. The Foundation has been able to check the degradation of mangroves in this vast stretch of land by providing complete protection, which has increased the forest cover of this region by 15 per cent during the last decade, which is commendable.

Chembur: Due to reclamation of coastal areas for housing as well as for industrial development, most of the mangrove areas have been lost from this region. Some of the remnants, however, show the presence of *Avicennia marina* with a few individuals of *Avicennia officinalis*. A solitary *Sonneratia alba* (about five year old) was observed growing near the Thane - New Mumbai creek bridge which is the result of the *Sonneratia* seeds carried ashore from the Elephanta island and is fairly a recent establishment.

Trombay to Sewri: The mangrove vegetation in this region is a mixture of *Avicennia marina, Acanthus ilicifolius, Sesuvium portulacastrum, Salvadora persica* and *Aeluropus lagopoides.* A few *Rhizophora mucronata* trees are also observed at Trombay. Major salt cultivation in Greater Bombay can be observed in this region. In the drier regions nearing salt pans, species such as *Suaeda fruticosa* and *Sesuvium portulacastrum* are observed while *Aeluropus lagopoides* is confined to the exposed banks of the water inlets that are frequented by people.

From Trombay to Sewri via Vadala *Avicennia marina* displays thick bushy growth. In some of the areas towards Sewri, small trees of *Avicennia marina* appear distantly scattered due to extensive harvesting for fuel. *Salvadora persica* makes its appearance on the drier patches alongwith *Aeluropus lagopoides* in the open.

Colaba: The entire coastal strip from Sewri to Colaba is devoid of mangrove vegetation. A few stands of *Avicennia marina* alongwith *Rhizophora mucronata* and *Sonneratia alba* are present at Colaba in the southern region of Mumbai island and extend up to Cuffe Parade on the west indicating their past prosperity.

Butcher and Elephanta Islands: Butcher and Elephanta islands are two major volcanic rock formations within the dockyard of Mumbai. Butcher island is managed by the Mumbai Port Trust. *Avicennia marina* present along parts of this coast, is subjected to the oil spills from tanker washes. The oil is also carried with tides to the Elephanta island, which is located just opposite Butcher island. This has resulted in heavy mortality of mangrove seedlings, which are observed growing naturally during rainy season. The deposition of oil on the pnuematophores of *Sonneratia alba* has also affected their growth.

Backbay Reclamation to Worli: The coast along the Backbay reclamation does not bear any mangrove vegetation except a small patch of *Avicennia marina* (about an acre or so) near Cuffe Parade. Human interference can be easily noticed in this region as almost half the wetland area has been reclaimed over the past few decades and the remaining coastal vegetation has been put under constant stress. The region from Malabar hills to Worli consists of huge rocks along the shore and a sandy beach from Worli to Mahim. *Avicennia marina* was growing sparsely in the region behind the Beach Candy Baths and Haji Ali's tomb in 1940s, but does not exist any more (Navalkar, 1941).

Mahim and Bandra: From Mahim bay as one enters the Mahim river towards the east, *Avicennia marina* occurs abundantly along the fringes of the estuary. *Rhizophora mucronata* and *Kandelia candel* are rare and threatened. Other species from this region are *Bruguiera cylindrica, Ceriops* tagal and a few stands of *Acanthus ilicifolius.*

From Mahim towards Bandra along the railway lines, *Avicennia marina* occurs with a few *Bruguiera cylindrica* and *Clerodendrum* bushes towards the drier soils. *Sesuvium portulacastrum, Suaeda fruticosa* and *Aeluropus lagopoides* are also present. From Mahim causeway along the shores, *Avicennia marina* occurs as bushes. From the Bay to Bandra point and further to Versova, some stunted *Avicennia marina* plants are still thriving, especially near the Band Stand, despite sandy nature of the coast. *Lumnitzera racemosa*, formerly a common species of this region in late 40s (Navalkar, 1941) is facing the threat of extinction, thanks to the highly polluted waters of the creek due to industrial growth of the Megapolis.

Madh: Opposite Versova along the protected eastern coast of Madh island, a few *Avicennia marina* and *Sesuvium portulacastrum* patches are observed alongwith *Acanthus ilicifolius* and *Aeluropus lagopoides.*

Malad: Malad creek with its mouth at Versova has large tracts of mud, however, besides a few scattered *Avicennia marina* and *Aeluropus lagopoides*, there is hardly any mangrove vegetation left. As one enters the creek towards upstream region, a few *Excoecaria agallocha* are seen intermingled with Avicennia marina and *Acanthus ilicifolius. Rhizophora mucronata,* a common species of this creek once upon a time (till mid-eighties), is no longer observed here.

Manori: As one enters Manori creek, the fringes are dominated by *Avicennia* stands and towards the village it appears alongwith *Aeluropus lagopoides* in the drier areas. The *Avicennia* stands are seen scattered along railway line almost upto the Mira road station. *Ceriops tagal, Acanthus ilicifolius* and *Salvadora persica* are some of the other species.

The most extensive tidal area (about 100 m in width) on the western side of northern Mumbai is provided by the low lying areas of the north, bordering Vasai and Malad creeks with their extensions and the Manori creek with its ramifications. These tidal areas interlace former paddy fields. The growth of the western suburbs up to Bhayandar and further north along the western railway has brought about a massive incursion in these areas. Here, coastal villages have been transformed in to beach resorts and residential colonies.

CURRENT STATUS OF MANGROVES

Distribution of mangrove vegetation of Mumbai and its surrounding metropolitan regions is chiefly composed of *Avicennia marina* which seems to be very prolific in growth and germination. It is only in a few places such as Ghodbunder where the intensity of human interference is not very marked, that we find a clear zonational pattern of mangroves like *Rhizophora mucronata, Bruguiera gymnorrhiza, Aegiceras corniculatum, Excoecaria agallocha*, and *Clerodendrum inerme*. The other species are *Avicennia marina, Avicennia officinalis, Sonneratia apetala, Sesuvium portulacastrum, Salicornia brachiata, Suaeda fruticosa, Kandelia candel, Salvadora persica, Derris heterophylla, Bruguiera cylindrica* and *Aeluropus lagopoides*; and they do not show any clear zonation as seen in Ghodbunder.

As far as the current status of the mangrove vegetation of Mumbai is concerned, various species observed in different locations in Mumbai, as reported by earlier researchers such as Blatter (1905), Navalkar (1939), Navalkar and Bharucha (1941) and Deshmukh (1990), have either disappeared from these locations or are under severe threat of extinction. Some of these species recorded during the current study are highlighted in Table 6.2.

Table 6.2: Current Status of Various Mangrove Species in Mumbai

Name of species	Earlier Records	Current status
Bruguiera parviflora	1905, Blatter (in Mumbai)	Disappeared
Carapa obovata	1905, Blatter (in Mumbai)	Disappeared
Kandelia candel	1905 (Blatter), 1941 (Navalkar & Bharucha) in Bandra & Andheri	Disappeared from these regions
Bruguiera gymnorrhiza	1905 (Blatter), 1939 (Navalkar) in Mahim	Disappeared from Mahim, rare in Ghodbunder
Lumnitzera racemosa	1939 (Navalkar), in Bandra	Disappeared from Bandra
Rhizophora mucronata	1939 (Navalkar), in Bandra and Andheri	Disappeared from these areas
Sonneratia caseolaris	1939 (Navalkar in Mumbai); 1987 (Lattoo) and 1990 (Deshmukh), in Elephanta	No more observed in Mumbai

CAUSES OF DEGRADATION AND LOSS OF COASTAL FLORA

In any given area, the physical presence of people exerts some amount of pressure on resources and this pressure increases with an increasing population (Aswani Kumar and Chaphekar, 1985). The population of Greater Bombay, which according to the 1991 census was 12 million, will increase by about 3.7 million, to 15.7 million as against the projected figures of 13 million in the year 2001 (Metcalf and Eddy Inc., 1979).

With the expansion of city limits including the greater part of surrounding Thane and Raigad districts, the area of Mumbai metropolitan region has increased. This, alongwith the tremendous increase in the population, the problem of habitation (housing) has become more acute resulting in reclamation of low lying areas for "development". Areas such as Mahim, Bandra, Andheri, Malad, Dahisar, Thane, Mulund, Bhandup and Chembur many of which were previously supporting luxuriant mangrove vegetation are now cleared and reclaimed to meet the growing demands of population for housing as well as for sites for new industrial units.

Despite the above pressure, it is heartening to see some of the luxuriant patches of mangrove forests in some areas. They are Vikhroli (Pirojshanagar), estuarine regions of Ulhas river, and Mahim in Mumbai and those at Mumbra and Karanja in the metropolitan region of Mumbai. Mangroves of Pirojshanagar (Vikhroli) and Karanja are already protected as they are under the ownership of private sector organisations and could serve as repositories for conserving mangroves by development of field level gene banks for posterity.

CONCLUSION

Problems affecting mangrove vegetation in Mumbai and its metropolitan region are manifold. Reclamation of land from swampy areas for setting up of industries and housing is a major one contributing to the disappearance of mangrove forests at an alarming rate. Dredging of sand from creeks has affected mangroves in areas such as Nagla Bunder, Mumbra, as well as near the mouth of Dharamtar creek, while decimation of mangrove shrubs and trees for fuel is common almost everywhere. Oil spills from tanker washes contribute to heavy mortality of mangrove seedlings at Sewri and Elephanta island (Lattoo, 1987). Though one would like to confirm cause and effect

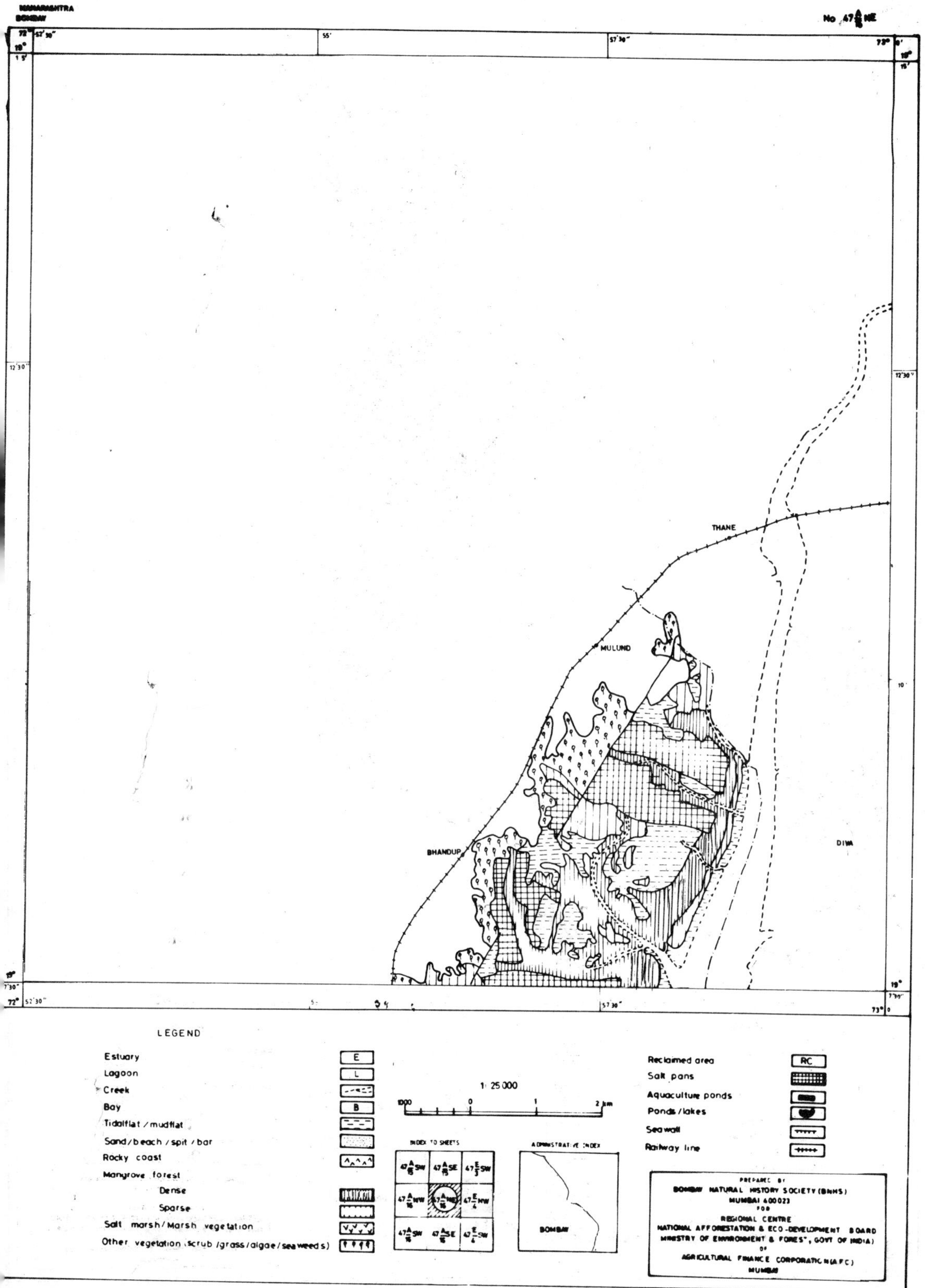

Figure 6.1: Coastal Wetland Map of Mumbai

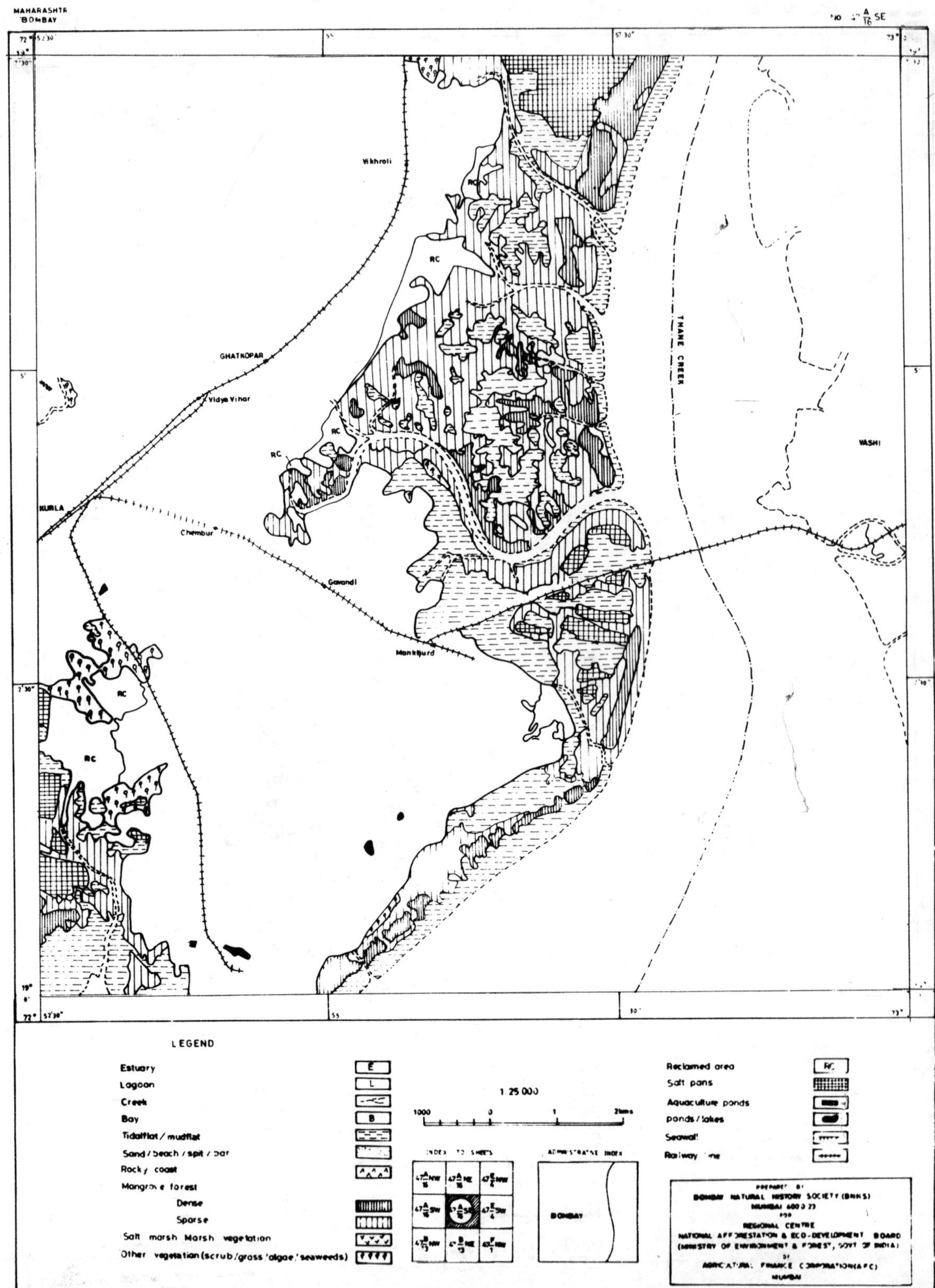

Figure 6.2: Coastal Wetland Map of Mumbai

relationships in these cases, we should not wait for all the necessary information to be made available before corrective measures are taken to control the loss of mangroves.

The remote sensing studies and the field surveys undertaken during this study highlighted an important fact that the Mumbai coast, despite the population pressure, still supports mangrove vegetation. Identification of appropriate areas in Mumbai and its metropolitan region for *in situ* conservation as shown below, could be one solution for preserving important gene pools of mangrove forests for posterity.

Areas proposed for *in situ* conservation are upstream regions (Ghodbunder upto Thane) of Ulhas river estuary, Mumbra mangroves (Thane district), Pirojshanagar (Vikhroli) and Mahim Mangroves (Mahim creek) (figures 6.1 and 6.2).

Rapid rate of disappearance of mangroves, especially on the account of changes in the land use, requires urgent attention and efforts have to be directed to protect representative areas for long-term studies and research.

ACKNOWLEDGEMENTS: We are grateful to the Regional Centre (Western Zone) - National Afforestation and Eco-development Board of the Ministry of Environment and Forests, Government of India for kindly providing financial assistance for undertaking the current study.

REFERENCES

Aswani Kumar, A.V. and S.B. Chaphekar. 1985. Ecology of estuaries: a case study of Ulhas river. *Perspectives in Environmental Botany* 1:67-85.

Blasco, F. 1975. *Mangroves of India*. French Insatiate, Pondicherry. 175 p.

Deshmukh, S.V. 1990. *Ecological Studies of mangroves in Bombay*. Ph.D. Thesis, Bombay University, Mumbai.

Deshmukh Sanjay, 1997. Final report of the project "*Eco-development of Mumbai coast: community based conservation and regeneration of mangrove forests: a case study*" submitted by BNHS to the Regional Centre (Western Region) of the National Afforestation and Eco-development Board (RC-NAEB), Ministry of Environment and Forests, Govt. of India. 121 p.

Lattoo, C.S. 1987. *Ecological observations on the vegetation of Elephanta island*. Ph.D. thesis, University of Bombay, Mumbai.

Maharashtra State Gazetteer. 1964. *Kolaba district*. Directorate of Government Printing. Stationary and Publication, Maharashtra State, Bombay, India.

Metcalf and Eddy Inc. 1979. *Report to the Municipal Corporation of Greater Bombay upon methods for treatment and disposal of waste water from Greater Bombay*. Metcalf and Eddy Inc. and Environmental Engineering Consultants joint venture. Pages 6-95.

Navalkar, B.S. 1941. Studies in the ecology of the mangroves of the Bombay and Salsette islands. Ph.D. Thesis, University of Bombay, Bombay.

USE OF THE OCEAN AS POLLUTION SINK IN MUMBAI

Rakesh Kumar

ABSTRACT

The Ocean is increasingly being used as the sink for the large volume of waste generated by the urban settlements situated near coastal areas. The west coast of Mumbai receives about 2225 million litres per day (mld) of sewage from the city with partial or no treatment that results in poor water quality. The expected rise in the flow by the year 2015 AD suggests that the water quality of the nearshore region would further deteriorate. The constraints of land availability in Mumbai has led to utilising the natural purification capabilities of the sea for waste dispopsal. The sea can easily assimilate the waste if discharged in the deep water with a long outfall. The study of Mumbai waste water scheme suggests that a combination of long outfall and primary treatment on land would be able to achieve the desired objective of better water quality of shore and efficient waste water management for the city of Mumbai. This Chapter presents the pros and cons of the waste management practices for a large metro city.

INTRODUCTION

The ocean being a vast natural sink with its potential to assimilate the waste, wastewater discharge in the ocean is a natural management option world wide, for all the coastal cities. In view of the large assimilative capacity in the oceans, it is also logical to use oceans for wastewater discharges. However, the problem occurs when anthropogenic activities create large and concentrated wastewater discharges in a limited area instead of dispersing them over large areas and at longer distances away from the coasts.

For a metro city like Mumbai, which generates about 2225 mld of sewage from seven service areas the problem of disposal can be complex. Of seven service areas, five of them directly or indirectly discharge massive wastewater load. Of the total current flows, more than 70 percent of the flow is generated from these service areas alone.

The development of wastewater management scenario for Mumbai warrants consideration of the natural sink in the ocean. The feasibility of using this sink requires proper design of wastewater treatment and disposal plan so that the natural assimilative capacity of the ocean is not depleted and there are no significant impacts on the flora and fauna of West coast.

The following sections delineate the present situation with regard to the Brihanmumbai Municipal Corporation plan for wastewater management, existing status of the west coast of the Mumbai region in terms of water quality attributes, expected improvements with the planned implementation of projects, situation of no action scenario by the year 2015 AD, pros and cons of various management options with regard to land availability, cost of land treatment and outfalls. Details have also been presented with regard to general pattern of waste treatment and disposal options for coastal cities in the world.

BOMBAY SEWAGE DISPOSAL PROJECT

The first integrated wastewater management scheme for the city was envisaged way back in 1970. The scheme included the option of disposal of wastewater through marine outfalls at Colaba, Worli and Bandra. A 1.1 km long outfall at Colaba is operational which is located in the harbour region.

The present plan envisages that Worli and Bandra will have preliminary treatment (screening and degritting) and 3 km long outfall for sewage disposal on the west coast and preliminary treated wastewater (by aerated lagoons) from Malad and Versova service areas will be discharged into Malad creek and from Ghatkopar and Bhandup service areas into Thane creek.

Despite full treatment in aerated lagoons for Malad and Versova wastewater, the receiving water quality criteria for the Malad creek discharges will be violated. These findings were reported in NEERI's report of 1994, entitled as "Environmental

Management Plan for Mumbai Sewage Disposal Project, Malad and Versova Aerated Lagoons". Based on the above findings, it was necessary that other options for treatment and disposal locations be explored for Malad and Versova service areas.

WATER QUALITY STATUS OF THE WEST COAST

The present wastewater discharge on the west coast have created many critical areas where water quality is severely impaired in terms of physico-chemical and bacterial attributes. High concentrations of pollutants are witnessed during the ebb tide compared to the flood tide. The Dissolved Oxygen (DO- an indicator of water quality with regard to organic pollution) levels show that about 5 km length of sea on the west coast is highly polluted, mainly the regions such as Malad and Mahim, which receive about 50 percent of the wastewater discharge from the city. The other 8 km stretch of the coast shows moderate depletion of the Dissolved Oxygen. The DO profile in the whole of west coast of Mumbai is presented in Table 7.1.

Table 7.1: Dissolved Oxygen Profile of the West Coast

Location	Dissolved Oxygen ranges, mg/l
Malad, Mahim	< 2 during ebb tide
Worli, Dadar and Bandra	2-4 during ebb tide
Girgaon, Juhu, Manori	> 4 during the ebb / flood tide

The other parameters such as Biochemical Oxygen Demand (BOD) and Nutrient levels are moderate in most of the regions. The sediment quality did not show any significant accumulation of metals and nutrients in the polluted region. The inorganic pollution such as metals in the marine water and sediments on the west coast is insignificant

Bacterial contamination in the west coast of Mumbai is high with levels ranging from 10^4 -10^5 counts/100 ml for Total and Fecal Coliforms. Only about 25 percent of the west coast (of a total 24 km length) has bacterial counts of Total/Fecal coliforms below 1000/ 100 ml.

Phytoplankton population shows prevalence of Cyanophycea in all regions where direct wastewater discharges are meeting the shore. Large population and high prevalence of Cyanophyceae is an indicator of high organic population.

Water quality in the Malad creek where the present Malad and Versova service areas are discharging the wastewater is extremely poor both in terms of physico-chemical and bacterial attributes. The BOD levels up to 40 mg/l, absence of DO and bacterial counts beyond 10^6 /100 ml are very frequent. It has also very high levels of ammonical nitrogen in the water and high total nitrogen in the sediments. Biological flora and fauna is almost absent in the Malad creek.

IMPLEMENTATION OF PRESENT SCHEME

Worli and Bandra

The implementation of the two outfalls at Worli and Bandra would improve the existing physico-chemical water quality considerably at all the impacted regions in these regions in these services areas. The improvement would be in terms of higher DO levels, negligible odour nuisance, very low BOD and also low nutrient levels near coastal zones and beaches. The values of DO would normally be more than 4 mg/l and BOD values less than 3 mg/l at the diffuser locations.

Sediment quality is not likely to be affected significantly due to the outfalls operation at Worli and Bandra. The earlier study (Environmental Management Plan for Bombay Sewage Disposal Project, Marine Outfalls, October 1993, by NEERI) of the areas in the vicinity of the existing outfall at Colaba indicated that no impairment of the sediment took place due to the years of operation of the 1.1 km long outfall.

Bacterial water quality at the shore is likely to improve with the two outfalls due to wastewater discharges 3 km away from the shore. The impact of the two outfalls at Worli and Bandra on the overall reduction in bacterial density at shore would be felt only if the non-point sources of the discharge presently reaching the shore line are eliminated. This is due to the fact that about 30-40 percent of the wastewater on the West coast is reaching directly to the shore. The general improvement in the water quality near shore will improve the fish spawning and breeding grounds.

Malad and Versova

Although the first phase wastewater facilities at Malad and Versova would comply with the stipulated effluent standards of discharge into the tidal creeks, these would not improve the Malad creek water quality. During the low tide, BOD levels of more than 20 mg/l, absence of DO and more than 10^5 coliforms/ 100 ml. would be frequently encountered. Even high level of treatment at Malad and Versova, or partial diversion of the flows, would not be sufficient to improve the water quality in the Malad Creek. This is due to the fact that Malad creek is a very small tidally influenced creek and has very limited assimilative capacity vis-à-vis the waste it receives.

For Malad and Versova discharges, the flows in the year 2015 AD is estimated to be about two times of the flows in year 2001 AD. Simulation of the wastewater discharge scenario for the Malad creek using the two dimensional model DIVAST was conducted with the flows of the year 2001 AD. Various management options for Malad and Versova were examined such as the diversion of flow, discharge into the Malad creek during the study, Environmental Management Plan for Mumbai Sewage Disposal Project, 1994. All these options indicated no significant improvements in the creek water quality despite secondary treated effluent discharge in the creek.

LAND TREATMENT VIS-À-VIS OUTFALL DISPOSAL

The conventional approach for any waste treatment for years has been always for land treatment. The urban pressure and the natural growth of densely populated area near coastal zone have raised new issues with regard to its merits and demerits. The alternative available for coastal cities in terms of long outfall sea disposal has shown promise worldwide.

The use of ocean outfalls for natural purification and assimilation for the domestic effluents has been very well documented. This simple process involves pretreatment for removal of floating material and large suspended solid (grit) before release of the wastewater at the sea bed with adequate water column. In a favourable situation when the wind, wave and current conditions are suitable, it provide high reduction in the BOD and bacterial density at a lesser cost and lesser damage to the marine environment than the conventional land based treatment plan (Simos et al., 1922).

It has been established that a properly designed outfall can provide an efficient and secure mechanism for the elimination of the environmental impacts due to sewage discharges in the marine ecosystem. The initial dilution provided by the outfall to the organic and nutrients reduces the concentrations to innocuous levels avoiding any ecological damage to the ocean ecosystem. It is possible that in many cases, the introduction of such substances to a usually nutrient deficient marine environment would probably be beneficial. An overall qualitative comparison between marine outfall and land based treatment methods is presented in Table 7.2 (Winslow and Moxan, 1928).

Table 7.2: Qualitative Comparison of Land Based Treatment and Outfall Disposal

Criterion	Most Favoured Inland Treatment	Alternative Marine Treatment
Acceptability of principle of treatment	*	
Compliance with mandatory water quality criteria		*
Compliance with mandatory discharges standards	*	
Minimum energy requirement		*
Maximum operational reliability		*
Safety to Public health	*	*
Least visual intrusion		*
Best odour control ability		*
Best noise control ability		*
Minimum area of land required		*

* Favoured

The concept of considering the land based secondary treatment as full treatmènt and outfall disposal as no treatment is misleading. The main difference between the two is the manner in which the overall reduction in the waste load is achieved. A secondary treatment reduces the load by controlled process regime, whereas in the ocean the natural processes carryout the similar reduction over a larger area. The efficiency of an outfall depends on the choice of the location of the outfall, design of the diffuser, initial dilution and the rate of loading. In a well managed system, presence of an outfall can lead to higher productivity of the commercial fishing as well (Cabellı et al., 1983).

SUITABILITY OF LAND BASED TREATMENT

Often it is assumed that the land based treatment provides highly pure effluent with very low pollution potential. It has been seen that a sewage treatment plant effluent is generally very rich in the nutrient contents as also the organic matter which get disagreed either in the sea or river. Natural decay and organic detritus from algal blooms can pose serious environmental changes and aesthetic sense could be affected just as badly be the decaying algae as by the crude sewage as demonstrated by Wilson (Streeter, 1951).

The efficiency of the secondary treatment does not indicate a health trend when it would be discharging the effluents with less than 30 mg/l of suspended solids and about $1x10^5/100$ ml of fecal coliform. A typical scenario of what would happen with regard to the hot spots of pollution in Mumbai, if effluents are discharged after secondary treatment in the near shore region is presented in Table 7.3.

Table 7.3: The Impacted Zone with Low DO (0-2 mg/l) and High Bacterial Density (1000-10000) near the Location of Discharge

Service Areas (flows in mld, 2015 AD)	Area Impacted in the Near Shore Region
Worli (728)	3-4 km
Bandra (1246)	5-6 km
Malad (1037)	whole Malad Creek + 2 km Length
Versova (433)	whole Malad Creek + 2 km Length

- The calculations are based on the present length of the affected region and increased flows of the year 2015 AD
- It assumes that Malad and Versova service areas would discharges in Malad creek after secondary treatment and the additional 2 km length in the seaward region beyond the creek mouth would also be affected.

Issue of Land Availability

The land requirement for each of the service areas is presented in Table 7.4. It may be possible to construct primary treatment facilities but the available space at Worli and Bandra may not be sufficient to locate secondary treatment. In Malad and Versova zones the land may be available for primary as well as secondary. However, the vacant land is comprising of mainly rich mangrove forests. A rich ecosystem of mangroves in Malad and Versova area should not be destroyed for secondary treatment purposes.

The consideration must always be given to factors related to land availability, land use, odour, and sludge disposal. Secondary treatment can acquire large area (about 0.052 ha/mld) and generally in a coastal area, the treatment site has to be near the coast to minimize the pumping requirement. Land in such areas are either not available and if available, the cost of the land is prohibitive. Besides, the aesthetics and odour problem that land treatment creates in the near shore region. Odour is probably a minor issue, but is case of sewage treatment plant, location within the 0.25 km could unacceptable odour levels and it could be unpleasant in the possible radius of 0.5-2.0 km depending upon the direction of wind.

Table 7.4: Land Requirement for Preliminary, Primary and Secondary Treatment (ha)

Service Areas	Preliminary	Primary	Secondary*
Worli	0.30	5.4	36.0
Bandra	0.40	14.0	64.0
Malad	0.32	11.2	51.2
Versova	0.15	2.7	18.0

* Secondary treatment also includes the preliminary and primary land requirements.

A further analysis also shows that operation and maintenance costs of marine outfalls are negligible compared to a full fledged secondary treatment as presented in Table 7.5. As is evident from the table the capital as well as operation and maintenance (O&M) costs of primary treatment and 3 km outfall are less than a full scale secondary treatment. Land requirement for outfall is comparatively low as the headworks could be located in a very limited space. Odour problem due to the sewage treatment works is completely eliminated.

Table 7.5: Capital and O&M Cost of Primary and Secondary Treatment and Outfall

Service	Primary Treatment		Secondary Treatment		Only 3 km Outfall		Primary + 3Km Outfall	
	Capital Cost	O&M Cost	Capital Cost	O&M Cost	Capital Cost	O&M Cost	Capital Cost	O&M Cost
Worli	9780	524.35	43452.8	1288.5	5010	50.1	14790	574.26
Bandra	16891	897.42	57796.2	2205.42	5010	50.1	21901	947.22
					Only 4 km Outfall		Primary + 4 Km Outfall	
Worli	9780	524.35	43452.8	1288.5	8330	83.3	18110	607.46
Bandra	16891	897.42	57796.2	2205.42	8330	83.3	25221	98.42
					Only 5 km Outfall		Primary+ 5 Km Outfall	
Worli	9780	524.35	43452.8	1288.5	1000	100	19780	624.16
Bandra	16891	897.42	57796.2	2205.42	1000	100	26981	997.12

* All values expressed in Million Rs.

CONCLUSIONS

Land based treatment upto secondary level are better for those areas where the sewage has to be discharged in the low assimilative capacity water body such as rivers, lakes and estuary. However, for larger agglomerate of the population situated near coastal areas, a properly desinged outfall combined with primary treatment at land can be more pragmatic solution at a comparatively low cost. In a situation, where providing safe water supply to the population is considered more important than catering to the need of waste disposal, best practicable solutions shold be used to derive the maximum benefits without hampering the quality of the marine water.

REFERENCES

Cabelli V.J., Dufour A.P., McCabe L.J. and Levin M.A. (1983): A marine recreational water quality criterion consistent with indicator concepts and risk analysis J.W.P.C.F. 55(10) : 1306-1314

Salas H.J. (1986): History and Application of microbiological water quality standards in the marine environment. Water Sci. Tech. Vol. 18, No.11

Scott.W.J. (1951): Sanitary study of shore bathing waters. Bull Hyg. 33 :351.

Simos G.W., Hilscher R., Ferguson H.F. and Gage S de M. (1922): Report of the Committee of Bathing Places Amer. J.Pub. Health 12(1) :: 121- 123.

Streeter H.W. (1951): Bacterial quality objectives for the Ohio river ; A guide for the evaluation of sanitary conditions of waters used for potable supplies and recreational use Cincinnati Ohio River Valley Water Sanitation Commission.

Winslow C.E.A. and Moxan D. (1928): Bacterial pollution of bathing beaches in New Haven Harbour. Amm. J.Hyg. 8 : 299.

SOME ENVIRONMENTAL AND SOCIO-ECONOMIC ISSUES OF MUMBAI

Vikas Tondwalkar and V. K. Phatak

INTRODUCTION

In the 16th century, Mumbai was a cluster of seven fishing islands which have now become a single land mass, in the form of a narrow peninsula. In 1840, Mumbai island covered an area of 18 sq. miles and population of about half a million. The colonial powers recognised the importance of protected harbours offered by these islands and developed it as a major harbour. Mumbai's transformation from a trading town into an industrial and manufacturing centre began when in 1854 the first chimney stack spewed black smoke onto the town's skyline. Since then Mumbai is growing, expanding, urbanising, as well as commercializing and today it has become the major financial and business capital of India.

Now the city of Mumbai with its western and eastern suburbs forms the Brihanmumbai covering an area of 468.07 sq.km. and a population of over 10 million people. In 2011, Mumbai's population is expected to become around 13 million. At present, Mumbai is the world's sixth largest city and by the turn of the century, it is expected to become the world's third largest city.

Geography and Climate

The climate of Mumbai is fairly equable since seasonal fluctuations of temperature are not significantly large. The moderating effects of nearby sea and fairly high amount of relative humidity in the atmosphere have restricted the variability. The prevailing winds are from north-west to south-west in sunshine and monsoon months whereas from the north in winter month. The temperature inversions do occur but last for only a few hours.

Natural Settings

In terms of land use, Mumbai's 87 sq. km. is covered by forests, 37 sq. km. is covered by agriculture and plantations, 11.5 sq. km. is covered by beaches and 99 sq. km. is coastal wetland area including 65 sq.km. of mangrove forests; but the largest part (209 sq. km) is now urbanised with residential, commercial and industrial uses (MEIP, 1994). Being an island, Mumbai has a long coastal line that consists of sandy beaches, exposed rocks, marshes, mud-flats, mangroves and salt lands. The coastline has undergone major changes over the past several decades because of the reclamation of wetlands for urban expansion, agriculture, salt pans, fish farms and also because of dumping of solid waste (MMRDA, 1994).

ENVIRONMENTAL AND SOCIO-ECONOMIC ISSUES

Mumbai is facing an urgent crisis in terms of its environmental sustainability and quality of life. Like other coastal cities of the world, Mumbai also has a potential threat from the global warming effect which may cause polar ice to melt, oceans to rise, and coastal lands and cities to flood. The effects of ozone depletion on the living beings, the dumping of toxic and hazardous waste into sea and the occurrence of acid rains and volcanoes have not been documented so far in Mumbai. Also there has not been any major environmental hazards due to earthquakes in Mumbai. Environmental problems of Mumbai are mainly on account of inadequate infrastructure and municipal services, congestion, lack of housing, inadequate public transport facilities and increasing number of vehicles. These are therefore, closely linked with social and economic aspects of development.

The major environmental and socio-economic problems of Mumbai can be summarized as follows:

Environmental Issues

a. Air pollution mainly from transport and refuse burning;
b. Noise pollution mainly from transport;
c. Poor services of refuse collection and disposal;
d. Degradation of soils, wetlands and mangroves because of urbanization;
e. Inadequate and contaminated water supply in some parts of Mumbai;
f. Poor sanitation and sewerage systems and
g. Degradation of Coastal Waters.

Socio-Economic Issues

a. Inadequate housing,
b. Inadequate,
c. Health hazards and
d. Congestion and crowding.

ENVIRONMENTAL ISSUES

Air Pollution and Noise Pollution

Transport sector has been identified as one of the major contributors to the total air and noise (MEIP, 1994; NEERI, 1995; URBAIR, 1996) pollution in Mumbai. The number of vehicles in Mumbai has been increasing at an alarming rate as presented in Table 8.1. The registered population of vehicles (NEERI, 1995) in Mumbai in 1995 is over 6.5 million apart from number of other vehicles which enter in the city from areas outside Mumbai. In Mumbai north-south railway corridors are served by Western and Central railways. The Western Railway operate about 923 suburban services per week day whereas Central Railway operate about 1072 services per week day.

A comprehensive emission inventory was developed in Mumbai for 1992-93 as a part of URBAIR project and the summary of which is presented in Table 8.2. From the emission inventory it is seen that the large number of emissions are because of the combustion from industry and power plants, followed by vehicular emissions and re-suspension from roads.

Table 8.1: Number of Vehicles in Mumbai (in Lakhs*)

Types	Year		
	1991	1992	1995
Two wheelers	242008	252301	275942
Three wheelers	24577	31278	46556
Cars	299289	287163	253560
Trucks	6501	4171	19322
Other Utility Vehicles	56086	71794	62535
Total	628461	646707	657915

* 10 Lakhs = 1 million.

Table 8.2: Summary of Annual Emissions in Mumbai, 1992-93 (Tonnes Per Year)

Emission Sources	TSP	PM10	SO2	NOx	Total (TSP+SO2+NOx)
*Transport sector (Diesel vehicles)	3673	3673	3490	19520	26683
*Resuspension from roads and others	10200	2500			10200
Fuel combustion from industries and Power Plant (Furnace oil)	3317	2996	64710	15285	83312
*Stone crusher	6053				6053
*Domestic/Commercial Sector (wood burning)	4432	2235	1688	1344	7464
*Refuse burning	4108	4108	26	153	4287
*Construction Marine (Docks) (Furnace Oil)	560	469	9350	1245	11624
TOTAL :	32343	15981	79264	37547	149623

* These sources exhaust emissions at low height and contribute to population exposure directly.

In addition to the vehicular emissions, resuspension from roads and the burning of domestic wood and garbage are among the largest contributors to air pollution problems in Mumbai because these sources exhaust emissions at low height and significantly contribute to population exposure. The emissions from industries do not contribute much to ground level exposure due to their tall stacks and because of favourable meteorological conditions. In Mumbai, the major problem of air pollution is from smoke and dust which largely contains Total Suspended Particles (TSP) and Particulate Matter less than 10 Microns (PM_{10}). The TSP emissions in Mumbai are estimated at about 32,343 tonnes per year for 1992-93 and out of this about 29,000 tonnes per, year are emitted at low heights containing about 22,000 tonnes per year of emissions of PM_{10}. Out of total SO_2 emissions of 79,764 tonnes per year in Mumbai, a major portion of about 64,710 tonnes is emitted from industries. It is also seen from the emission inventory that in Mumbai the thermal power plants and diesel vehicles contribute maximum NOx emissions.

Municipal Corporation of Greater Mumbai (MCGM) operates a large Ambient Air Quality Network comprising about 22 stations which are located in commercial, industrial and residential areas. These Stations regularly monitor Suspended Particulate Matter (SPM), sulphur dioxide (SO_2) and nitrogen dioxide (NO_2) which are of primary concern due to their effects on health. Ambient air quality data as monitored by MCGM for 1992 and 1994 is presented in Table 8.3

Ambient air quality results depicts that SO_2 values during 1992-1994 did not exceed the Central Pollution Control Board (CPCB) standards at all the locations monitored whereas for NOx the values at Sion, Maravali, Worli and Sewree exceeded the standards in 1992. However in 1994 the NOx values exceeded the standards at only Sion and Khar. The SPM values exceeded the standards in 1992 and 1994 at all the locations monitored except for Mankhurd and Kalbadevi areas.

The ambient Air Quality Monitoring (AQM) in Mumbai does not cover the monitoring of transport related pollutants such as CO, HC and lead. There is a need to strengthen and improve ambient air quality monitoring network of MCGM.

The ambient noise levels in Mumbai are increasing mainly due to growing road, rail and air traffic. Unlike air pollution noise does not accumulate in the environment and gets dissipated rapidly. However, noise can have adverse health impacts during the short period when its intensity is high. The average noise levels in different residential, commercial, industrial and silence areas of Mumbai in 1989 are presented in Table 8.4 (CPCB data). The ambient noise levels during the day and night times in residential, commercial and industrial areas of Mumbai and silence zones exceeded the permissible standards. The high noise levels in these areas can be attributed to road and rail movements. The residential, commercial and industrial activities are located on either side of these rail and road corridors thereby exposing the population to higher levels of noise pollution problems.

Table 8.3: Ambient Air-Quality Data for 1992-1994 (Annual Mean)

Station	SO2		NOx		SPM	
	1992	1994	1992	1994	1992	1994
Colaba	14	14	47	33	153	202
Worli	32	30	67	51	222	236
Dadar	22	24	56	43	225	230
Parel	30	29	74	52	302	324
Sewree	56	31	66	32	170	183
Sion	21	30	89	74	300	290
Khar	-	31	-	63	-	339
Andheri	30	37	53	40	253	276
Saki Naka	22	22	57	28	228	249
Jogeshwari	14	18	42	36	288	303
Ghatkopar	33	26	58	34	247	241
Bhandup	58	57	45	29	237	244
Mulund	33	33	40	29	225	252
Borivali	8	10	32	31	210	220
Chembur Naka	26	23	62	42	266	322
Maravli	15	19	70	53	296	420
Anik Nagar	26	33	47	40	207	239
Mankhurd	9	31	16	37	132	216
Kalbadevi	8	25	30	36	144	229
CPCB Standard for Residential areas.	60	60	60	60	140	140

Solid Waste Management

The collection of refuse from each and every part of Mumbai City is the most difficult and challenging task considering the huge quantity of refuse generated (about 5000 tons/day) over a vast area as well as because of narrow roads and difficult access in slums. Therefore, adoption of number of variants of collection methods is necessary in different parts of the city to achieve adequate level of service. A short fall in collection efficiency may have an adverse impact in city's environment resulting in unhygienic conditions, spread of diseases and aesthetic impairment. In Mumbai, except in slums and congested areas most of the solid waste generated is adequately collected by Municipal Corporation. However, in slums and other congested areas which are located in fringe areas and swampy locations, the refuse mostly remains uncollected and part of which finds its way into storm water drains.

Table 8.4: Average Noise Levels in Different Areas of Mumbai in 1989

Area	Period	Noise Level in dB(A)				CPCB std
		Leq	L90	L50	L10	
Residential	Day time	70	65	69	76	55
	Night time	62	51	54	64	45
Commercial	Day time	75	69	74	77	65
	Night time	66	57	65	72	55
Industrial	Day time	76	69	75	80	74
	Night time	65	60	64	67	70
Silence Zone	Day time	69	58	66	71	50
	Night time	60	48	54	62	40

The refuse collection in Mumbai is predominantly by community bins served by a variety of vehicles which carry the collected refuse directly to disposal sites or to transfer stations. Some localities in Mumbai are also served by house to house collection system. The various types of vehicles used for transportation of refuse include compactors, dumpers placers, dumpers, trucks, jeeps ,tractors, metadors and bulk refuse carriers(NEERI, 1995).

The solid waste disposal practices followed in Mumbai are quite primitive and environmentally unsatisfactory. The four currently operated municipal dumping sites at Deonar, Gorai, Malad and Mulund are reclaimed mangroves and creek areas that have been acquired by landfilling with refuse and debris. Leachates and refuse from these dump sites find their way into creeks and may cause serious environmental hazards (MEIP, 1994; NEERI, 1995)

In addition to this, during dry months widespread burning of refuse is allowed at dumpsites which creates severe air pollution problems in the localities situated in down wind directions. Manual scavenging by informal waste pickers occurs substantially in Mumbai and the health problems associated with this are unknown. Also study on Environmental Management Strategy (EMS) for MMR conducted under Metropolitan Environment Improvement Programme (MEIP) has indicated that toxic and hazardous wastes generated in Mumbai are not recognised as such, and therefore accepted for co-disposal at the municipal dumping sites.

Degradation of Soils, Wetlands and Mangroves

As a result of land filling, reclamation and dredging the land forms and soils of Mumbai have been substantially modified (MEIP,1994). Soils are contaminated due to numerous industries discharging large quantities of toxic heavy metals in the environment. The concentrations of heavy metals in soils has shown enrichment factors ranging for lead from 0.98 to 1.98, for Zinc from 0.93 to 1.94, for cadmium from 6 to 8.79, for Copper from 0.7 to 2.35 and for Nickel 1.26 to 3.33 when compared with the rural background samples(MEIP,1995).

Land filling operations and reclamation in various parts of Mumbai have been conducted to provide sites for housing, industries, highways, refuse dumping etc. which has resulted in destruction of mangroves and wetlands along with the development of certain low lying flood prone areas and creation of storm water drainage problems. In addition to public and private agencies that are responsible for the wetlands and mangroves degradation, certain slum dwellers that depend upon them for high calorific fuel and food have added to the degradation. The loss of wetlands and mangroves is expected to result in loss of ecological productivity in terms of filtration ability of wetlands, biotic productivity, and breeding grounds for insects and pathogens as well as loss of area of natural drainage, deprivation of open spaces, the loss of flood absorption medium and aesthetic quality.

Water Losses and Pollution Problems

EMS study has reported (MEIP, 1994) that the usable water is lost because of the following reasons. Also, water pollution problems are on rise.

- **Surface Evaporation from Reservoirs**: In case of lakes supplying water to Mumbai, about 30% of the volume of water held in lakes is lost in evaporation.
- **Leakages and pipe bursts:** In case of leakage detection programme in MCGM, about 23% of water supply is estimated to be lost because of leakage from the pipelines.
- **Avoidable wastage:** In slum localities as water supply is very limited, the residents create artificial leaks or tamper with the pipelines to obtain additional supplies. Also, overflowing of water from the overhead tanks of multi-storied buildings during peak hours results in large quantities of water wastage. To control such water losses substantial efforts from both local authorities as well as general public are required.

The discharge of untreated domestic and industrial wastes has deteriorated the water quality in all the creeks and most of the stretches of coastal waters in Mumbai. Also beach waters are characterized by high turbidity which exceeded 30 NTU and high bacterial contamination. The beaches at Juhu and Dadar were observed to be more polluted during summer than in winter. There has been no uniformity and regularity in water quality monitoring in Mumbai.

SOCIO-ECONOMIC ISSUES

Economy

The economy of Mumbai has witnessed structural changes during the past decade which are reflected in higher contribution of the tertiary sector activities (trade and services) in Mumbai's income and employment as compared to secondary sector (manufacturing activities). The income of secondary sector has grown at 3.9% while the tertiary sector has grown at 6% per annum during 1980-89. Similarly, the share of manufacturing sector employment has reduced from 36% in 1980 to 28.5% in 1990. Whereas, in trade and services sector, it has increased from 52.1% to 64.3%. Significant contribution to Mumbai's economy due to increase in trade and services related activities and decrease in manufacturing activities, has resulted in prevailing infrastructure and service related environmental problems, rather than industrial pollution problems. The economic growth rate of Mumbai has been significantly higher as compared to Maharashtra because of its most diversified and vibrant economy.

Congestion and Crowding

In addition to the presence of slums, inadequate infrastructure and health problems, the general public of Mumbai is facing the problem of increased pedestrian and vehicular conflict at most of the railway station areas and commercial aeas. Because of the presence of narrow roads, large number of temporary stalls set up by hawkers selling consumer goods on the footpaths, presence of rickshaw stands, bus depots etc. near the station areas, along with large number of pedestrian and traffic movements, these places have become heavily congested during peak hours. This results in related environmental problems such as increased air and noise pollution caused by idling and slow speed vehicular movement, slow pedestrian movement as well as risk of accidents.

Environmental Assessment Study (MMRDA, 1997) being conducted for proposed flyover at Khodadad Circle under MUTP-II has indicated that the foot paths in this area are congested with 400 temporary stalls set up by hawkers and each stall owner earns about Rs.3000 to 4000 per month estimating economic turn over of about Rs.15 to 16 lakhs per month.

Health

The environmental health status in Mumbai is largely affected by the fact that half of its population lives in slums or on pavements. The health problems are prevalent due to congestion, inadequate water supplies, lack of garbage clearance and disposal systems, lack of proper housing and ventilation, existence of slums, traffic and noise problems, industrial pollution and also because of ignorance and carelessness about the medical treatment (MEIP, 1994; NEERI, 1995) The principal health problems as a result of poor sanitation and poor living condition in slums are gastroenteric infections, helminthic infections, infectious jaundice, Malaria and Filaria. A study conducted by the Tata Institute of Social Sciences(TISS, 1992) under MEIP in 1992 which examined the environmental status of six slums in Mumbai had indicated the health problems in slums are acute and the majority of slums and pavement dwellers suffered from diarrhoea, dysentry, typhoid, bronchitis, jaundice as well as Malaria and filaria. In slums most of the people are so engaged in the business of survival that they do not consider health problems to be of great priority. In Mumbai although there are government and municipal hospitals nearby, majority of population prefers to be treated by private practitioners in spite of heavy charges as in Goverment health centres.

In addition to the health problems due to poor sanitation and living conditions, the rising levels of air and noise pollution are becoming a major public health concern. Studies conducted by Environmental Pollution Research Centre at KEM Hospital Studies (KEM, 1993), Mumbai has indicated that the excess mortality due to PM_{10} (and TSP) was about 2,765 cases and of an exposed population of 9.8 million. An estimated impact of PM_{10} on health in Mumbai in terms of morbidity is indicated in Table 8.5.

Table 8.5: Estimated Impact of PM_{10} Air Pollution on Health in Mumbai, 1991.

Type of health impact	Number of cases (thousands)
Chronic bronchitis	20
Restricted activity days	18680
Emergency room visits	76
Bronchitis in children	190
Asthma	741
Respiratory symptom days	60(millions)
Respiratory hospital admissions	4

The population exposure to TSP is the major air pollution problem in Mumbai, which may have adverse impact on human health. The studies (URBAIR, 1996) conducted on population exposure for annual average TSP in Mumbai have shown that the road side residents, public transport drivers, traffic police and other roadside workers (about 3% of the population) and resident near stone crushers are the most seriously exposed to TSP. This study has also indicated that about 97% of the population exposed to TSP concentration above WHO standards (90g / m3) and about 8% of the population to that of more than twice the WHO standards.

The health status of the general public in industrial areas of Mumbai especially Chembur has been reported to be badly affected considering the high incidences of respiratory diseases, headaches and eye irritations. Increased noise levels in Mumbai due to traffic induced noise may cause health problems such as headache, sleeplessness, restlessness etc. in the nearby residential areas (MMRDA, 1992).

RESPONSES

In order to achieve sustained environmental development, the implementation of appropriate preventive and mitigative measures to stop any further environmental degradation is necessary. The measures adopted for environmental improvement should not however jeopardize the basic objective of higher economic growth as both economy and environment in turn contribute to better quality of life. Authorities of Mumbai have responded to various environmental problems and following measures have been taken up/suggested.

Control of Air and Noise Pollution

The abatement measures for controlling deterioration of air and noise quality should address in improving traffic and transportation situation in Mumbai. The following technical and policy measures should be given priority for mitigating air pollution problems.

Traffic Related Problems

- Introduction of clean fuel technology by oil and petroleum industries.
  a. Improvement in diesel quality by reducing sulfur content, by addition of detergents and changing physico-chemical properties would reduce PM_{10} emissions (modification to India's energy policy).
  b. Introduction of unleaded gasoline.
  c. Introduction of low smoke lubricating oil for two stroke engine will substantially reduce the PM_{10} from traffic exhaust.
- Adoption of clean vehicle emission standards by oil industries, vehicle manufacturers and importers as well as car owners.
  a. Fitting of exhaust gas control devices to 4 stroke gasoline engines and open loop catalysts to 2 stroke gasoline engines would control VOC, PM_{10} & CO emissions.
  b. Using of tailpipe emission treatment to diesel vehicles and existing buses retrofitting with equipment will reduce emissions.
- Improving traffic management by a variety of measures such as traffic control, traffic lights, new roads, area traffic control sytems etc.

Strengthening ambient air quality monitoring network in terms of continuous long term monitoring, revised monitoring network, increased pollutant coverage to account for traffic pollutants and also standardized monitoring methodologies.

Refuse Burning and Domestic Emissions

Extension of pubic refuse collection system to avoid open refuse burning and improvement and modernization of solid waste disposal system is necessary. Also refuse burning at the disposal sites should be stoped. To avoid wood burning, encouraging use of electric crematoria, and encouraging bakeries and commercial establishments to switch over to clean fuels.

Industrial Emissions

As the combustion of furnace oil from industries and power plants is the main source of PM_{10}, SO_2 & NO_x emissions, alternative fuel should be used.

Noise Pollution Problems

Traffic management measures used for control of air pollution also would help in mitigating noise pollution due to transportation activities. The following additional measures could be employed for control of noise pollution in Mumbai :

- Introduction of modern technology for railways to mitigate noise pollution due to movement of railway traffic.
- Constructing of barriers along the sensitive land uses such as schools, hospital, etc.
- Improved road dividers, pavements and implementations of stringent vehicle parking norms to reduce excessive honking.
- Introduction of concrete policy measures to control 'social noise' generated due to public processions, festivals, marriages, parties, political meetings and such other activities.

Improvement in Solid Waste Management System

For improving overall solid waste management in Mumbai, EMS study and Solid Waste Management in Greater Mumbai (NEERI, 1994) conducted under Metropolitan Environment Impovement Programme have recommended following measures.

- There is a need to assess the feasibility for establishing joint ventures and privatizing the different parts of refuse management systems in Mumbai.
- Development of methods/programmes for waste minimization, organized rag picking, recycling of waste is necessary.
- Implementation of a dedicated vehicle repair and maintenance and upgradation programme should be done.

- Introduction of regulation banning the collection and disposal of Toxic and Hazardous Waste & municipal refuse.
- Introduction of regulation banning disposal of waste in wetlands, mangrove or other areas of ecological significance.
- Identification and implementation of suitable disposal sites and alternative disposal methods with required environmental protection measures.
- Develop & conduct education & awareness programmes on waste minimization, litter prevention and civic responsibility for clean streets, community bins and drains.

Improvement of Sanitation and Sewerage System

EMS Study (MEIP, 1994) conducted under MEIP has recommended following measures for improving sanitation and sewerage system.

- Unsewered slums and non-slum areas of Mumbai should be covered by underground sewerage system in a phased manner,
- Areas of cross connections between storm sewers and fossil sewers should be identified and rectified,
- Provide low cost community toilet facilities for slums, pavements dwellers and along the railway lines to avoid open air defecation,
- Municipal sewage treatment before discharging it into creeks or sea, and
- Targeting on reuse of treated effluent for industrial or other purposes.

Protection of Soils, Wetlands and Mangroves

The measures for protection of soils, wetlands and mangroves are summarised below:

- Strengthen the existing soil conservation policy to better manage and protect natural resources through improved irrigation practices, environment friendly farming techniques, afforestation, control of industrial discharges, etc.
- Formulation, implementation and enforcement of a wetland/mangrove conservation policy to resolve the conflicts with regard to salt pans, solid waste dump sites, new roads, reclamation etc.

- Plantation of mangrove species in degraded wetland areas of Mumbai should be immediately carried out.
- Control of encroachment into wetland and mangrove areas is a must, which can be achieved by planning a buffer zone to avoid encroachment.

MMRDA'S INITIATIVE

MMRDA has prepared a long term perspective plan for the entire metropolitan region. This plan attempts to integrate various sectoral policies like industries, office location, shelter, water resource development, transport, etc. in an environmentally sustainable manner. It also provides a framework for integrating development control and environmental concerns through the mechanism of Environmental Impact Assessment.

MMRDA has been playing a crucial role in Environmental Management of MMR by carrying out strategic environmental planning, coordinating development and focusing attention on various environmental protection and heritage conservation issues in the Region. MMRDA has taken up another important step by setting up two separate Societies for Environmental Improvement and Heritage Conservation. The MMRDA has decided to provide Rs. 50 million as initial corpus to each of the Societies. The interest earned on this initial corpus shall be utilized by the Societies for providing financial assistance in the form of grants to all the societies, institutions, universities, colleges, Non-Government Organizations (NGOs) and voluntary Organizations, etc. in MMR for projects related to environmental improvement and heritage conservation.

However, improving and protecting the environmental status of Mumbai is not only the responsibility of Government alone. Efforts of NGOs, Voluntary Organisations, Industrial Organisations, Academic Institutions, Social Workers and Individual Citizens are equally important.

REFERENCES

KEM, (1993):. "Summary of studies done by Mahashur A.A., Dalal N. & Gregrat J.A of Environmetal Pollution Research Centre - KEM Hospital, 1993.

MEIP, (1994): "Study on Environmental Management Strategy and Action Plan for Mumbai Metropolitan Region", Vol.III, Curent Environmental Status in BMR, Prepared by Coopers & Lybrand, U.K. and Associated Industrial Consultants (I) Pvt.Ltd. India for Government of Maharashtra under Metropolitan Environmental Improvement Programme, April, 1994.

MMRDA, (1992):"Environmental Assessment of The proposed Jogeshwari-Vikhroli Link, Draft Final Report" prepared for MMRDA by AIC Watson Consultants Ltd. 1992.

MMRDA, (1994): "Draft Regional Plan for Mumbai Metropolitan Region 1991-2011" prepared by Planning Division, Mumbai Metropolitan Region Development Authority, October, 1994.

MMRDA, (1997): "Environmental Assessment of the proposed flyover at Khodadad Circle, Draft Final Report" prepared for MMRDA by AIC Watson Consultants Limited, Mumbai, 1997.

NEERI, (1994): Soild Waste Manaegment in Greater Bombay, Prepared by National Environmental Engineering Research Institute, 1994.

NEERI, (1995):"Draft Environmental Status of Greater Mumbai "prepared for Municipal Corporation of Greater Mumbai by National Environmental Engineering Research Institute, 1995.

TISS, (1992): "A study on Urban Environmental Management through Community and NGO" prepared under Metropolitan Environment Improvement Programme by Tata Institute of Social Sciences, 1992.

TOI, (1995): News from the Times of India 29th June 1995 quoting office of the Transport Commissioner.

URBAIR, (1996): "Urban Air Quality Management Strategy in Asia", URBAIR Greater Mumbai Report prepared by Norwegian Institute for Air Research, Norway, Virje Universitiet, Netherlands and Aditya Environmental Services India, October, 1996.

APPLICATION OF LANDSAT - TM DATA FOR MARINE POLLUTION ASSESSMENT OF MUMBAI COAST

G. K. Tripathy

ABSTRACT

In this Chapter, the use of satellite remote sensing approach to detect the water quality parameters has been discussed. A case study on Mumbai coast using latest approach is described. The Landsat - Thematic Mapper (TM) data in the wavelength range of 0.45-0.90 μm was applied to detect and measure the turbidity of marine pollution in the coastal region of Mumbai, known as Thane Creek, and its surrounding fresh water lakes. Collection of marine water samples in the Thane Creek region was synchronised with the Landsat- TM passing time. The water samples were analysed for turbidity and a relationship was established between the real time values of turbidity and the radiance values from the remotely sensed data. The results were verified with the three fresh water lakes adjoining the coast including one receiving the domestic sewage in the suburbs of Mumbai. The techniques could be utilised to detect the source and quantity of pollution in other coastal cities of India.

INTRODUCTION

The density of population in India is highest along the coastal towns and cities of the country because of the location of main industrial and commercial centres. There are inherent problems associated with coastal wetlands and marine bio-diversity due to increasing human interference. It may be due to either deforestation, agricultural conversion, hydrologic alteration, water quality degradation, ground water depletion, species extinction or domestic sewage disposal and discharge of chemical wastes.

India has around 7% of world's mangroves spread over seven states. The present estimate of the total coverage is 4240 km^2(Govt. of India, 1991) which means that 67% of the original mangrove cover has been lost since the pre-agricultural times (Anon, 1991). The loss of coastal wetland forests reduce the ability of wetlands to slow down water movement and trap suspended sediments. Chilka lake, the largest brackish water lake of the country, is being filled with suspended sediments coming from the deforested upland erosion. The area of Chilka lake has been reduced from 860 km^2 to 605 km^2 in the last 60 years (Malini et. al., 1993). In the coastal stretch of India's busiest city, Mumbai, the disposal of chemical and organic wastes from the chemical factories and domestic sewage have increased the BOD and COD values in the water. Though there has been lot of speculation about the increase of mangroves in this coastal stretch but decline in annual fish catch is observed by the city's fishermen. The reduction in fish population may be due to the extinction of mangrove species in the locality and increase in BOD and COD in the sea water. The increasing coastal pollution poses threat to the marine ecosystem. The awareness programme on wetland issues and consequences of failed conservation efforts calls for renewed attempts with more effective methods of tracking the coastal problems. The latest techniques of remote sensing applications could be the answer to some of these.

Satellite Remote Sensing (RS)

Remote sensing implies sensing an object from a distance or remotely without being in actual physical contact with it. It involves the measurement of electromagnetic energy (EME) reflected or emitted from the objects on the surface of earth. Most satellite sensors operate in visible (0.4-0.7 μm) and Infra Red (IR) (0.7 - >3 μm) portions of the spectrum. Depending on the applications of the satellite these ranges of spectrum called bandwidth are selected for choosing the bands. The surface features have different reflection properties in these spectral bands.

Spectral Properties of water

Remote sensing (satellite, microwave) activities take place in the spectral range of 0.45 - 12.5 μm. The EME reflected from the surface of water is being sensed by the satellite sensor as digital count. The amount of EME reflected or emitted will depend on the characteristics properties of the water. Bartolucci et al. (1977) have shown that the spectral response variation of turbid and clear water varies between the range 0.5μm to 0.95μm. The range affected by turbidity are mostly bands 1 through 4 of the Landsat TM (0.45-0.52μm, 0.52-0.60μm, 0.63-0.69 μm, 0.76-0.90μm). It is noted that the high concentrations of phytoplankton in clear ocean waters have lower reflectances than water in the range of 0.45-0.51μm, but higher reflectances for all wavelengths above 0.51μm and extending into the near-infrared. The range affected by turbidity and chlorophyll are approximately replicated by bands 1, 2, 3 and 4 i.e., a range between 0.45 μm to 0.90 μm of the Landsat TM.

RS to Coastal Pollution Detection

Of all the civilian satellite data, mostly used satellite data for the identification and quantification of coastal pollution is the Landsat TM. Indian Remote Sensing (IRS) satellite series IRS 1A, 1B, 1C data are also being processed to track the turbidity and chlorophyll contents of the water. Indian satellite IRS - 1C has got better appreciation over Landsat TM because of its improved spatial resolution (23.5 m) or the data quality and the similarity of IRS bands with those of Landsat TM. The bands of IRS series (0.45-0.52 μm, 0.52- 0.59 μm, 0.62- 0.68μm, 0.77-0.86 μm, 1.55- 1.70 μm) have got the best use of applying to track the coastal pollution. Band 2 is being used directly locating the source and relative concentration of turbidity.

Remote sensing technique has got advantages over the traditional methods due to its time and cost saving method, accuracy and temporal and spatial observations on water features. Though many ways of digital data transformations and applications to quantify the coastal pollution have been evolved but the success criteria is yet to be realised. The influence of sea ambience has several effects on the reflection properties. In addition, the water characteristics upto a depth of 20 metres can only be determined by any space borne satellite sensor. On the other hand, the underlying sediments and organic wastes from the sea floor are the constant source of sea water pollution. Measurement of turbidity, total suspended solids and the chlorophyll have been very successful in the recent days due to introduction of improved sensor resolution, coastal zone colour scanner and lately the microwave radar satellites. Identification of oil slick, mangrove species identification, growth of coral reefs and fishing zone potentials are some of the other applications of remote sensing where its usefulness are best realised.

REVIEW OF EARLIER WORKS

There are many research outcomes on water pollution described by several authors in the peer reviewed scientific journals. Most of them have tried out the study using Landsat TM (LS-TM) data. Lathrop and Lillesand (1986) used Landsat TM scenes to estimate pollution parameters such as turbidity, chlorophyll and suspended solids for the Green Bay and Lake Michigan study areas. Ground reference data of water samples for these parameters were acquired nearly coincident with the LS-TM over pass and then analysed in the laboratory. TM data in all the first 4 bands (1 through 4) were first georeferenced and then atmospherically corrected followed by extraction of 3×3, 5×5, 7×7, 9×9 pixel windows corresponding to the ground water sample point. The digital data in all the 4 bands were converted to radiance ($mW.cm^{-2}.sr^{-1}$) so as to maintain the uniform scale. Linear regression method was used to quantify the relationships between the various water quality parameters and the radiance values of TM bands. The multiple correlation coefficient (R^2), the standard error of the mean Y estimate, F-values and biasness tests were used to establish the statistical significance of the regression models. Numerous regression models were evaluated and the best model determined for each water quality parameter was used to prepare the pollution image depicting that parameter throughout these two study areas. Such type of study using SPOT satellite data has not been very successful except only the detection of turbidity of water in Band 2 data (Cairns et. al., 1997).

On the other hand, Forster et. al. (1993) used a similar kind of approach to a coastal sewage outfall site off Sydney Harbour, Australia for generating the pollution images of turbidity, chlorophyll -a and -b, phaeopigment and total pigment. Sampling of sea water was synchronised with the Landsat- TM pass over. Before correlating water sample data with the TM digital data, the four TM bands were transformed using the Chebyshev Series form. The series equations developed for each band were used as the independent variables in a standard multiple regression equation. The Chebyshev form is preferred since it leads to much better accuracy both in the computation of the coefficients and subsequent evaluation of the fitted polynomials. The regression equation for each water quality parameter was tested and the results applied to the full Landsat TM image in all the four bands.

Models have also been developed to determine various water quality parameters and quantitatively calculate their concentration and produce distribution maps (Aranuvachapun and LeBlond, 1981; Baban, 1993; Khoram, 1985; Tassan and Sturm, 1986; Jing, 1991). Environmental sensitive maps on sea coast can be prepared by merging all data through Geographic Information System (GIS) as suggested by Jing and Yan, (1991) and Allewjin, (1994).

THE CASE STUDY OF MUMBAI COAST

During the last decade, the Thane creek, adjoining Mumbai (Bombay) city, has become an important dumping site for treated effluents and unchecked pollutants from domestic as well as various industrial sources such as textile, petrochemical and the bulk chemical industries. This narrow strip of coastal water is very important from environmental point of view since it supports a vast area of mangrove forest besides a wide variety of flora and fauna. In view of this, it is of prime concern that the pollutants affecting marine water quality parameters and their spatial distribution should be properly understood and monitored.

In this connection the capabilities of satellite remote sensing in simultaneous coverage of larger areas at regular interval has come as a boon in discerning temporal and spectral heterogeneity of the marine pollution. Sea water sampling at Thane Creek was synchronised with the Landsat passes on 21st May 1994 (Behera et. al., 1996). The samples were collected from 6 locations and analysed in the laboratory with respect to turbidity and expressed in NTU(Nephelometric Turbidity Units). Results of turbidity measurement for the representative samples are shown in the Table 9.1.

Remotely Sensed Data Used

The digital data of Landsat Thematic Mapper of 21st May, 1994 were obtained. The data set covers the area of Thane Creek and adjoining water bodies such as Powai Lake, Vihar Lake, Tulsi Lake and a part of Ulhas River.

Table 9.1: Turbidity Analysis of Samples of Thane Creek

Locations	Turbidity in NTU
1	13
2	8
3	14
4	15
5	15
6	8

NTU- Nephelometric Turbidity Unit

Processing of Landsat TM Band Data

The digital band data formed 850 lines by 456 columns (flat raster format) in four spectral bands, 1 through 4. TM digital band data were geocoded with the toposheet to establish appropriate sampling positions. It was done in two steps, namely; i) geometric correction(GC) and ii) resampling (RS), with the help of DIPIX Interactive Image Processing System(IIPS). Geocoded band data were transported to Geographic Information System (GIS) environment where other ancillary information (i.e. water quality parameter) was available for statistical analysis.

Study of FCC Images

Two False Colour Composite (FCC) images of TM bands 1, 3 and 4 were developed. Both the images are the northern and southern extensions of each other respectively representing Thane Creek. Except for a plume noticed nothing much in contrast is observed in these images. The images were also developed for band-1 (the band mainly delineates turbidity) for both southern and northern parts of Thane Creek. Although a sharp contrast is observed between the northern and southern parts of Thane creek, the spatial distribution of turbidity is not clearly understood and most importantly, the quantitative assessment of turbidity cannot be explained through the simple digital numbers.

Model Development

Single TM pixel in the image is strong enough to relate the corresponding sample values on the surface. But for sea ambient environment (sea dynamics and hydraulic changes) and instrumental error, it was therefore considered appropriate to use an averaged pixel window value to improve the signal noise ratio (Foster et al., 1993). In the present study, an averaged pixel value of 3x3 window was used in model development. The three step development procedure of the model can be described as follows:

1. Chebyshev Polynomial Equations were developed for band data on treating with the turbidity data for each location.
2. Chebyshev approximated band data were regressed against the turbidity data set acting as the dependent variable.
3. Chcbyshcv approximated band data were regressed against the turbidity data in a multiple regression analysis where turbidity data set acting as the dependent variable.

The Chebyshev form is chosen for better accuracy over normally used least square method (Rivlin, 1974; Fox et al., 1968). The Chebyshev series is of the form :

$$f(X) = 0.5\, a_0\, T_0(X) + a_i\, T_i(X)$$

where $T_i(X)$ is the Chebyshev polynomial of the first kind of degree i.

Thus, for example,

$$T_0(X) = 1$$

$$T_1(X) = X$$

$$T_2(X) = 2(X)^2 - 1$$

$$T_3(X) = 4(X)^3 - 3(X)$$

where $X = (2x - x_{max} - x_{min}) / (x_{max} - x_{min})$, X is normalised to run from -1 to +1.

In this equation "x" represents Landsat TM band data (averaged through 3x3 array) acting as the independent variable, whereas f(X) represents turbidity data set acting as dependent variable. Four Chebyshev polynomial equations were generated for four TM band data. Each of these equations now form the independent variable against the turbidity data (dependent variable) in a multiple regression. The details about the theory of regression analysis is given in Draper et al. (1966) and Davis (1973).

A multiple regression equation is of the form:

$$Y = a_0 + a_i X_i + e$$

where X_i represents the output value of the Chebyshev Polynomial equation developed for TM band data and Y represents the turbidity data. The regression equation for the turbidity was tested and the results applied to the whole Landsat-TM image (bands 1, 2, 3 and 4) and turbidity images were generated for the entire scene.

Regression Model

Chebyshev polynomial equations were developed using 6 sampling sites. The degree of polynomial approximation was found to be 1 for all the four bands. The coefficients of Chebyshev transformation are given in Table 9.2. The polynomials were then used as independent variables for multiple regression with turbidity (Tu) as dependent variable. The multiple regression developed is as follows:

$$Tu = -107.377 + 0.033\ TM1 + 0.45\ TM2 - 0.124\ TM3 + 0.006\ TM4$$

where TM1, TM2, TM3 and TM4 are the mean count values in Landsat TM bands 1,2,3 and 4 respectively, and Tu is turbidity expressed as Nephelometric Turbidity Units. Multiple correlation coefficient (R) was calculated as R=0.98 with F-value=10.72 significant at 99% level of confidence (Table 9.3).

Table 9.2: Coefficients of Chebyshev Transformation

Bands	a_0	a_1	a_2	a_3
1	10.65	20.74	4.37	23.44
2	16.37	11.28	-0.1	-5.73
3	14.04	14.04	0.90	10.76
4	15.44	-20.81	-5.05	18.37

Note: a_0, a_1, a_2, a_3 are the coefficients of Chebyshev Transformation

Table 9.3: Multiple Regression of Chebyshev Transformed Turbidity Data of Bands 1,2,3,4

Source of Variation	Sum of Squares	Degree of Freedom	Mean Squares	F-test
Regression	49.67	4	12.42	10.7198*
Deviation	1.1586	1	1.1586	
Total Deviation	50.8286			

Regression Coefficients are a_0 (-107.377), a_1 (0.0331), a_2 (0.4497), a_3(-0.1235), a4 (0.0064)

* Significant at 95 % level of confidence.

Application of Model to Landsat TM Data

The regression model was applied to the Powai, Vihar, Tansa Lakes and the Ulhas River near Bombay. It is revealed that the turbidity level of Vihar Lake and Tulsi Lake(which is presently used for drinking water) is quite low, whereas turbidity is much more pronounced in Powai Lake and the Ulhas River. Powai Lake receives the domestic sewage from the nearby huge inhabitation and the Ulhas river receives a large quantity of industrial effluents as well as domestic sewage. The quantification of the turbidity is expressed in NTU.

CONCLUSIONS

The application of Chebyshev Polynomial equations and then multiple regression analysis was found to be a satisfactory method for characterisation and prediction of turbidity. The methodology justifies its validity for extrapolating it outside the study area as has been observed in case of Powai Lake, Vihar Lake, Tulsi Lake and a part of the Ulhas River. Secondly, it has got an advantage over simple False Colour Composite (FCC) and raw band data (in this case, band 1 data, which records characteristic spectral signatures of turbidity) image in spite of analysis of physical pollution parameters (here, turbidity). Nevertheless, this method has been tested in any other coastal cities but it is hoped that due to the encouraging results coming out of the present study, it should be extended to the other part of the country.

REFERENCES

Allewijn, R., (1994), Proceedings of the Second Thematic Conference on Remote Sensing for Marine & Coastal Environments, held in New Orleans, Louisiana,USA on 31Jan.-2Feb.1994 (Michigan:Environmental Research Institute of Michigan), vol.1, pp.1149.

Anon. (1991), World Resources 1991-1992, New York: Oxford University Press, pp.291.

Aranuvachapun, S. and LeBlond P.H.,(1981), *Remote Sensing of Environment*,vol.11, pp.11.

Baban, S. M. J., (1993), *International Journal of Remote Sensing*, vol.14, pp.1247.

Bartolucci, L. A., Robinson, B. F. and Silva, L. F., (1977), Photogrammetric Enng. and Remote Sensing, vol.43, pp.595.

Behera, B., Tripathy, G.K., Inamdar, A.B. and Asolekar,S. R. (1996), Assessment of Turbidity of Seawater Using Remote Sensing Data, vol.25, pp.103-108.

Cairns, S.H., Dickson, K.L. and Atkinson, S.F. (1997), An Examination of Measuring Selected Water Quality Trophic Indicators with SPOT Satellite HRV Data.

Davis, J. C., (1973), Statistical and data analysis in geology, John Wiley and Sons, New York, 2nd edn., pp.650.

Draper, N. R., and Smith, H., (1966), Applied regression analysis, John Wiley and Sons, N. York, pp.709.

Forster, B. C., Xingwei, S. and Baide, X., (1993), *International Journal of Remote Sensing*, vo.15, pp.2759.

Fox, L. and Parker, I. B., (1968), Chebyshev Polynomials in numerical analysis, Oxford University Press, London, pp.205.

Government of India, (1991), A Reference Manual. Research and Reference Division, Ministry of Information and Broadcasting, GOI, Delhi, India, pp.920.

Jing, L., (1991), *Asian- Pacific Remote Sensing Journal*, vol.3, pp.59.

Jing, L. and Yan, L., (1991), *Asian- Pacific Remote Sensing Journal*, vol.3, pp.75.

Khorram, S., (1985), *Photogrammetric Engg. and Remote Sensing*, vol.51, pp.53.

Lathrop, R.G. and Lillesand, T.M. (1986), Use of Thematic Mapper Data to Assess Water Quality in Green Bay and Central Lake Michigan, *Photogrammetry Engineering and Remote Sensing*, vol.52(5), pp.671-680.

Malini, B.H., Rao, K.S. and Rao, K.N., (1993), Evolution and Dynamics of the Chilka Lake. The Management of Coastal Lagoons and Enclosed Bays, pp.257-268, Proceedings of the 8th Symposium on Coastal and Wetland Management, New Orleans, Lousiana, America Society of Civil Engineers, New York, USA.

Rivlin, T. J., (1974), The Chebyshev Polynomials, John Wiley and Sons, N. York, pp.186.

Tassan, S, and Sturm, B., (1986), *International Journal of Remote Sensing*, vol.7, pp.643.

Lathrop, R.G. and Lillesand, T.M. [illegible] Use of Thematic Mapper Data to Assess Water Quality in Green Bay and Central Lake Michigan. *Photogrammetric Engineering & Remote Sensing*, vol. [illegible] no. [illegible] 671-680.

[illegible] H. [illegible] (1992) [illegible] The Management of [illegible] 1-26. Proceedings of the [illegible] Symposium [illegible] Coastal and [illegible] Orleans, Louisiana [illegible]

[illegible] The [illegible] pp. 18.

[illegible]

10

HEALTH STATUS OF THE COASTAL WATERS OF MUMBAI AND REGIONS AROUND

M.D. Zingde and K. Govindan

ABSTRACT

The coastal marine environment of Mumbai and regions around receives the domestic wastewater generated of around 1800 million litres per day (mld), mostly untreated. In addition, a large number of industries established in the drainage zones also contribute to pollution loads. These inputs have affected the water quality, sediment quality and biological characteristics of receiving waters to varying degrees. BOD in coastal water is often high and water is enriched in dissolved nitrogen and phosphorous compounds. Although the concentrations of Petroleum hydrocarbons (PHC) and phenols are high particularly in Thane creek, they are not alarming. The levels of heavy metals such as Cr, Cu, Zn, Pb, Cd and Hg have enhanced, though marginally. The degraded water quality of coastal waters is also reflected in very high counts of pathogenic bacteria with their populations markedly decrease as the flood tide progresses. Adsorption and transfer of pollutants to bed sediment by suspended particulate matter has increased the sediment burden of several pollutants in some areas. Degradation in water quality has induced marked alterations in community structure and population of phytoplankton, zooplankton and macrobenthos. The impact of anthropogenic pollutants decreases considerably from creek to offshore areas and is insignificant beyond 5 km from the shoreline.

INTRODUCTION

The coastal marine environment of Mumbai and regions around, comprising of Arabian Sea to the west and a number of tidal inlets such as Thane creek, Back bay, Mahim creek, Versova creek, Ulhas estuary and Bassein creek (Figure 10.1), is the recipient of a variety of pollutants emanating from domestic and industrial sources. Of these, Thane creek receives wastewaters from heavily industrialised Thane-Belapur belt stretching along its eastern shores, a group of industries largely concentrated around Chembur and domestic wastewater through several point sources stretching between Colaba and Thane. In addition, the industries located around Rasayani release the effluents into Patalganga estuary while those at Nagothane use Amba estuary for effluent disposal. Both these estuaries open into Dharamtar creek which is an inlet of Thane creek. Loading and unloading operations at Mumbai and Jawaharlal Nehru ports and ship-generated wastes including bilge also expected to contribute to the environmental degradation of Thane creek. Back bay, Mahim bay and Versova creek are the recipients of large volumes of domestic wastewaters apart from industries located within their drainage zones.

Several industrial complexes and estates located in Mumbra-Ambarnath segment release wastewater in the Ulhas estuary, directly or indirectly, which opens in the Arabian Sea through the Bassein creek. Salient features of the prevailing water quality, sediment quality and biological characteristics of the coastal marine environment of Mumbai is discussed in this chapter.

WASTEWATER FLUXES

Reliable estimates of quantities of wastewaters entering coastal waters of Mumbai are not available. Guesstimates vary between $1.5x10^6$ and $1.8x10^6$ m^3/day for domestic wastewater released often untreated or partially treated through point discharges in Thane, Mahim and Versova creeks, Back bay and directly into the Arabian Sea off Worli. The fluxes of various contaminants entering marine zone through domestic wastewater are presented in Table 10.1 based on the flow of $1.8x10^6$ m^3/d (Zingde, 1989). These loads are quite significant since they are largely confined to inshore creeks and bays, which due to restricted water exchange, have limited capacity to assimilate pollutants. Thus, for instance though tidal range of 2-4 m generally occurs along the open coast, a spring tide rise of 3 m in Mahim bay decreases to 0.9 m in the outer Mahim creek and becomes negligible a few kilometers inland. Hence, the wastewater released in Mahim creek tends to stagnate and tidal flushing is slow.

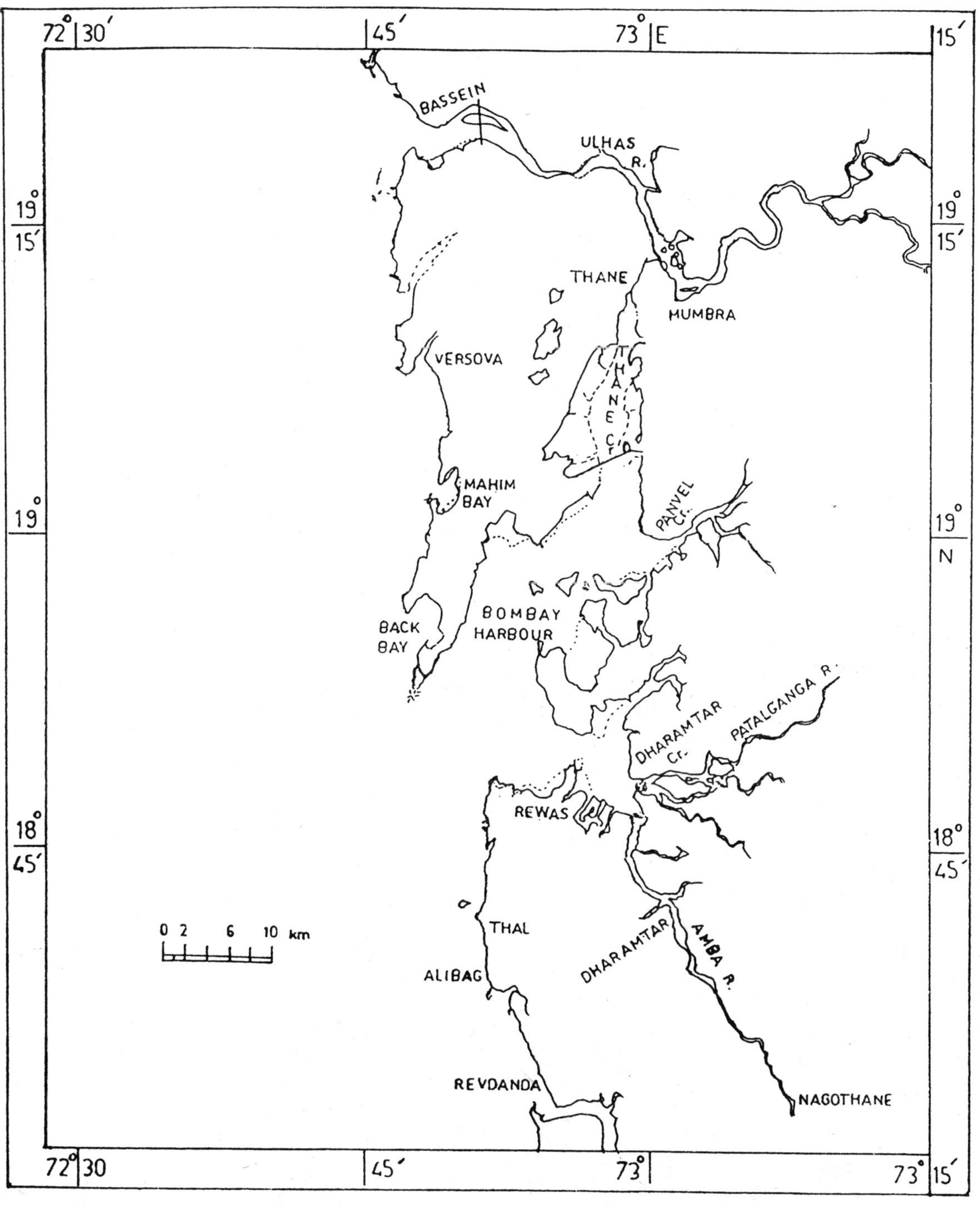

Figure 10.1: Coastal Environment of Mumbai

Table 10.1: Wastewater Characteristics and Loads of Contaminants Entering Marine Waters

Parameter	Average Concentration (mg/l)	Load (kg/d)
Dissolved solids	1500	2.7×10^6
Suspended solids	235	4.2×10^5
pH	7.0	-
BOD	250	4.5×10^5
COD	350	6.3×10^5
Total nitrogen	35	6.3×10^5
Total phosphorus	6	1.1×10^4
Oil and grease	7	1.3×10^4
Phenols	0.4	720
Chromium	0.02	36
Manganese	0.6	1080
Iron	2.0	3600
Cobalt	0.03	54
Nickel	0.05	90
Copper	0.08	144
Zinc	0.2	360
Lead	0.05	90
Cadmium	0.01	18

WATER QUALITY

Salinity

The salinity of the coastal water is under the influence of land drainage during monsoon though the impact is small beyond 10 m depth contour. During periods of high monsoon precipitation salinities as low as 1 ppt can occur in interior Thane creek, Mahim creek and Ulhas estuary. The high freshwater flow in these creeks enhances the flushing and the environmental quality improves considerably during monsoon. During dry season, however, the fluctuations in salinity in these waters are largely due to wastewater discharges. Thus, in Mahim bay the seawater salinity of 35 ppt intruding the creek during high tide falls below 15 ppt during low tide due to the influences of wastewater. The incomplete mixing of low saline, contaminated creek water and high saline sea water often results in appreciable vertical salinity gradient in some of these creeks. The offshore water however, is generally well-mixed vertically as well as laterally during the dry season.

Suspended solids

The strong tidal currents sweeping the shallow creeks disperse the fine grained and poorly sorted bed and shore sediments into the water column resulting in high suspended solids in water. The load though variable, is generally below 100 mg/l during dry season and increases substantially during monsoon (Zingde *et al* 1989). Off the open coast, the suspended load is generally below 50 mg/l (dry season) and is largely inorganic in nature with organic carbon content of less than 3%. It appears that the load of 4.3 x 10^5 kg/d of BOD associated with domestic wastewater does not contribute substantially to the suspended load in the coastal water though its contribution in Mahim and Versova creeks is high.

pH

Although the pH of openshore water (8.0-8.3) is close to that expected for seawater, a marked decrease occurs in the interior Thane, Mahim and Versova creeks and Ulhas estuary where the fluctuating pH often decreases to less than 7.5 around low tide under the influence of microbial oxidation of organic matter.

DO and BOD

DO in a water body is an important parameter since the existence of aquatic life including fishery is intimately linked with its availability. The sources of DO in water are photosynthesis and dissolution from the atmosphere across the air-water interface. However, the DO is consumed during microbial oxidation of organic detritus which is measured in terms of BOD. Some fraction of DO is also utilized by higher organisms through respiration. Although considerable controversy exists over the levels of DO required for a healthy ecosystem, it is considered that the DO should not fall below 2.5 mg/l for prolonged periods.

Release of large volumes of organic wastes influence DO levels in the creeks substantially as evident from Table 10.2. Generally DO increases with the incursion of tidal water and attains maximum around high tide followed by decrease during ebb with very low values around low tide. Extreme low values in Mahim and Versova creek throughout a tide cycle indicate the severity of pollution in these creeks. Even in Mahim bay where the tidal flushing is appreciable during spring tide, complete depletion occasionally occurs at low tide. The incomplete mixing of contaminated low saline creek water draining into Mahim bay with seawater, often results in steep vertical DO gradient and it is not surprising to observe bottom DO of as high as 2.5 ml/l when the DO at the surface is close to zero (DOD, 1992). DO in the open-shore water is generally above 3 mg/l but substantially undersaturated even upto 10 km from the shoreline indicating the influence of organic matter even in the offshore region (DOD, 1992). Unpolluted seawater has low BOD of generally less than 3 mg/l. The BOD in the creeks around Mumbai is variable and tide dependent and the values can exceed 20 mg/l in Mahim and Versova creeks. The open-shore water has considerable lower BOD though values as high as 4 mg/l are sometimes encountered.

Table 10.2: Typical DO at Selected Locations in Coastal Water of Mumbai During Premonsoon

Location	DO
Thane creek (Bhandup)	0.5-3
Thane creek (Colaba)	2-3
5 km off (Harbour mouth)	2-5
10 km off (Harbour mouth)	3-5
Mahim creek	0-0.5
Mahim bay	0-3
Versova creek (Inner)	0-1.5
Versova creek (Outer)	0-3
Ulhas estuary (Mumbra)	0.5-3
Ulhas estuary (Kasheli)	1.5-3.5
Ulhas estuary (Bassein)	2.5-4
5 km off Ulhas estuary mouth	2-5
10 km off Ulhas estuary mouth	3-5

Nutrients

Concentrations of PO_4-P, NO_3-N, NO_2-N, NH_4-N have increased considerably (DOD, 1992) in the creeks and the levels are often abnormal as compared to the concentrations in the offshore as evident from Table 10.3. The low levels of NO_3-N in Mahim creek and Versova creek can be explained on the basis of general mechanism of decomposition of organic matter. When the DO is depleted, the oxidation of organic matter proceeds utilising nitrate; the most efficient oxidant after DO. Hence, in the absence or in presence of low levels of DO, the nitrate entering the creeks during the high tide is consumed after the depletion of DO, leading to low NO_3-N. In the absence of DO, nitrate and nitrite, the oxidation of organic matter proceeds through the utilization of sulphate which is the abundantly available in seawater resulting in the liberation of sulphide and ammonia in Mahim and Versova creeks (Inner). Most marine organisms cannot survive under these conditions (Zingde and Sabnis, 1994).

Nitrite occurs in surface seawater in low concentrations as an intermediate product during the oxidation of ammonia to nitrate. It is evident from Table 10.3 that abnormally high NO_2-N occurs in the interior Thane creek and inner Ulhas estuary. This accumulation of NO_2-N in the system is indicative of an environment under high pollution stress.

Table 10.3: Typical Levels of Nutrients (μ mol/l) at Selected Locations in Coastal Water of Mumbai During Premonsoon.

Location	NO_3^--N	NO_2^--N	NH_4^+-N	PO_4^{3-}-P
Thane creek (Bhandup)	5-40	3-15	5-10	5-10
Thane creek (Colaba)	2-15	0.2-3	0.2-5	1.5-5
5 km off (Harbour mouth)	3-15	0.2-2	0.2-1	0.5-4
10 km off (Harbour mouth)	1-10	0.1-1.5	0.1-1	0.5-3
Mahim creek	0-5	0-10	10-80	5-25
Mahim bay	0.5-20	0.5-3	3-30	3-15
Versova creek (Inner)	0-2	0-5	15-50	10-70
Versova creek (Outer)	0.1-15	0.1-10	5-30	5-40
Ulhas estuary (Mumbra)	25-40	5-20	3-15	5-20
Ulhas estuary (Kasheli)	15-30	5-15	0.5-10	5-15
Ulhas estuary (Bassein)	10-25	0.5-3	0.1-3	3-10
5 km off Ulhas estuary mouth	5-15	0.1-1	0.1-2	0.5-2
10 km off Ulhas estuary mouth	1-10	0.1-1	0.1-1	0.5-2

Heavy metals

The available information indicates that the levels of dissolved Co, Ni, Cu, Zn, Pb and Hg are highly variable. The trends, however, suggest probable marginal increase in inshore waters as compared to that along the openshore (Sahu and Bhosale, 1991).

Petroleum Hydrocarbons

Although the inshore waters, particularly Thane creek receives high fluxes of petroleum contaminating wastes from refineries, tanker washings, loading and unloading operations of crude oil and petroleum products, the levels in water (1 m below surface) are relatively low (< 10 μg/l). It is possible that the PHC entering the water is scavanged by the high suspended load and removed from the water column.

SEDIMENT QUALITY

Pollutants such as trace metals, some organics, phosphorus etc. get adsorbed on the particulate suspended matter and transferred to the bed sediment on settling. Hence, the character of the bed sediment has changed considerably in inshore water bodies with considerable enrichment of certain pollutants in the surficial sediment. This is clearly evident from Figure 10.2 in which depthwise variation of trace metals, organic carbon (OC) and total phosphorus (TP) in a core from Mahim creek is illustrated. The surface concentration of Cu is as high as 800 ppm and that of Zn exceeds 2500 ppm as against 180 and 400 ppm respectively associated with the bottom segment of the core probably since the bottom sediment would have deposited when the creek was relatively free from pollution. Organic carbon, Total phosporous and Pb also reveal high accumulation. Similar trends have also been observed in cores obtained from certain locations from Thane creek though the metal accumulation is not to the extent expected considering the fluxes entering the system (Sharma *et al* 1994). It has been estimated that the top 20 cm sediment in Mahim creek contains 1.3 t Ni, 37 t Cu, 100 t Zn and 4.6 t Pb in excess over the expected background (Sabnis, 1984). Surficial sediment from Ulhas estuary also have higher burden of Cr, Zn, Pb and Hg (Bhosale and Sahu, 1991).

In Thane creek, the total Hg concentration in the surface sediment varies from 0.17 to 8.21 ppm with definite enrichment in the inner creek. An excess of 14 t Hg over the natural background has been estimated to be trapped in these sediments (Zingde and Desai, 1981). The sediment from Thane creek also reveal marked enrichment of PHC in surficial layers and values exceeding 3 μg/g (wet wt) are common. The contamination of open-shore sediment, however, is low.

BIOLOGICAL CHARACTERISTICS

Microbiology

Microbiological results (Table 10.4) indicate very high total viable count (TVC) in the creek systems of Versova, Mahim and Thane as compared to openshore waters (DOD, 1992). The population of total coliforms (TC) and many of the pathogens in these creeks exceed the required water quality standards and the bacterial contamination of the sea food is a distinct possibility since, these creeks serve as recognised landing centres for fish. Indiscriminate release of domestic wastewater is largely responsible for the prevailing microbial contamination of the coastal water.

Table 10.4: Populations (no/ml) of Some Microorganisms in Seawater Around Mumbai (Flood Tide Value in Parenthesis).

Bacteria type	Thane creek	Mahim bay	Versvova creek	Offshore
TVC	228000 (5600)	800000 (2900)	290000 (69000)	39
TC	9800 (210)	11000 (72)	9000 (79000)	6
ECLO	3200 (60)	6000 (26)	5600 (3600)	1
SHLO	2800 (110)	2000 (32)	3100 (400)	1
SLO	300 (-)	1000 (2)	600 (100)	-
PKLO	100 (40)	12000 (28)	1600 (-)	1
VLO	7600 (240)	1000 (56)	8200 (2300)	4
VPLO	2900 (30)	- (-)	400 (200)	-
VCLO	3900 (110)	200 (34)	5400 (1400)	-
PALO	4200 (70)	-- (21)	6600 (4200)	1
SFLO	6400 (70)	6000 (150)	8100 (7300)	2

TVC - Total viable counts; TC - Total coliforms; ECLO - *Escherichia coli* like organisms; SHLO - *Shigella* like organisms; SLO - *Salmonella* like organisms; PKLO - *Proteus klebsiella* like organisms; VLO - *Vibrio* like organisms; VPLO - *Vibrio parahaemolyticus* like organisms; VCLO - *Vibrio cholerae* like organisms; PALO - *Pseudomonas aerugenosa* like organisms; SFLO - *Streptococcus faecalis* like organisms.

Phytoplankton and phaeopigments

The creek systems, in general, reveal higher concentration of phaeopigments and phytoplankton counts as compared to openshore (Figure 10.3) due to higher nutrients availability associated with sewage pollution. However, the diversity of phytoplankton is relatively low as compared to the offshore coastal system. Mahim and Versova creeks

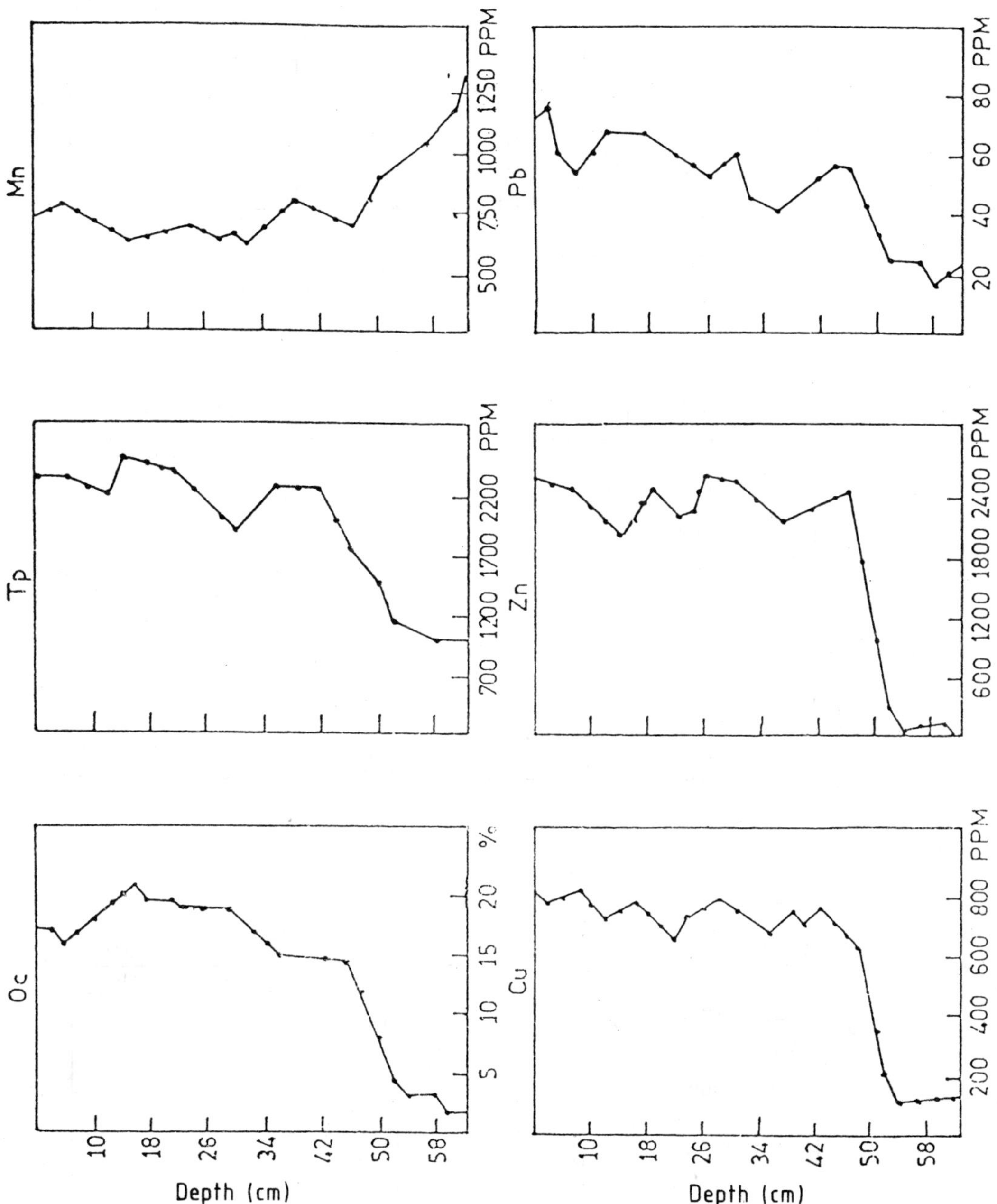

Figure 10.2: Depth Distribution of Organic C, P, Mn, Cu, Zn and Pb in the Core from Interior of Mahim Creek

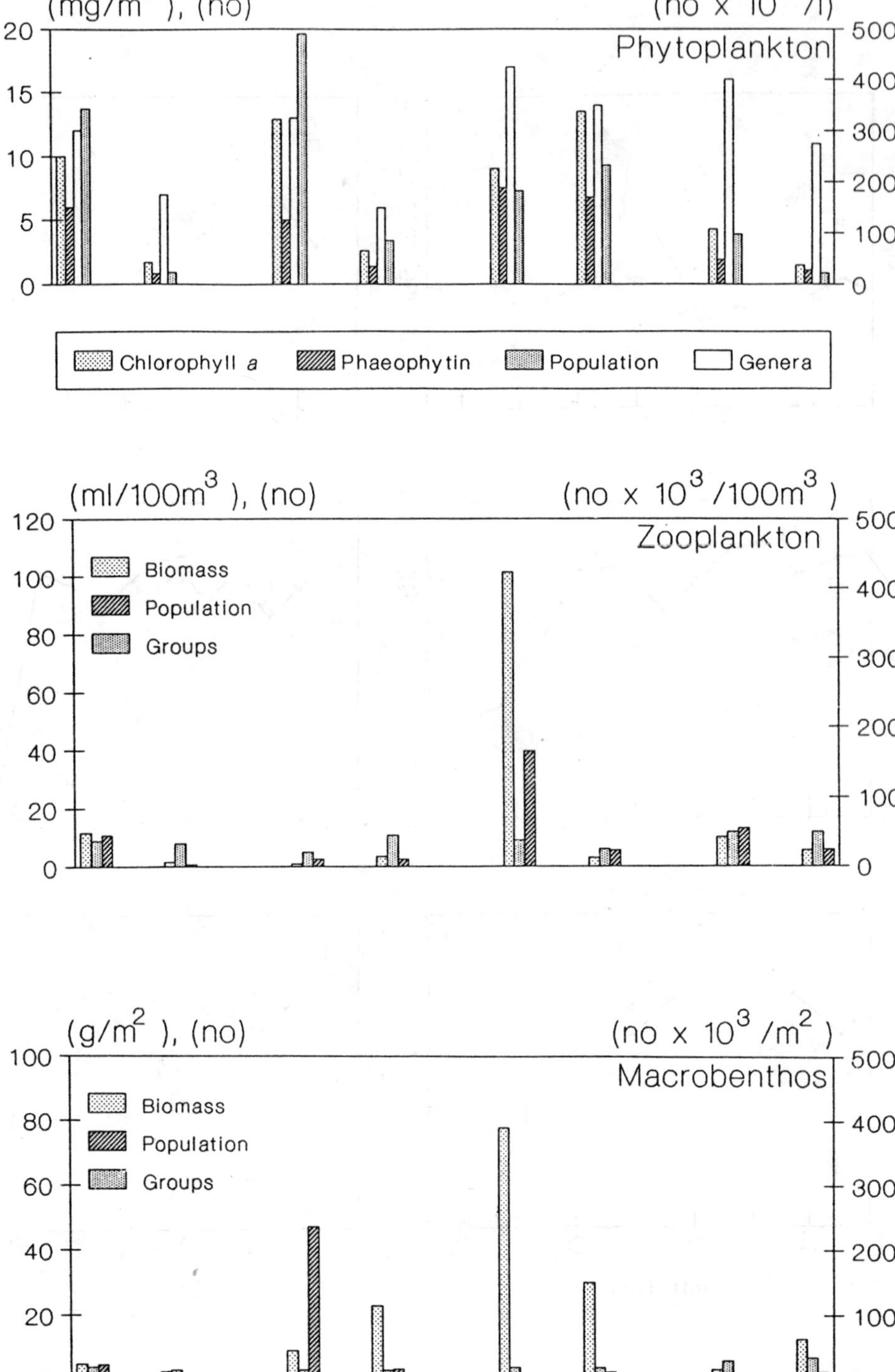

Figure 10.3: Comparison of Phaeopigments, Phytoplankton, Zooplankton and Macrobenthos between 1982-85 and 1995-96.

sustain relatively less diversity as compared to Thane creek. The trends indicate a reduction in the concentration of phaeopigments and phytoplankton counts and diversity, especially in Versova and Mahim creeks in recent years when compared with the past results (Figure 10.3). A change in generic dominance of phytoplankton (Table 10.5) between 1982 and 1996 is also evident (Ram, 1985; DOD, 1996). This could be due to the response of tolerant species to the change in pollution stress. However, Thane creek shows an increasing trend in phaeopigment values and phytoplankton counts during recent years than the past.

Table 10.5: Composition of Phytoplankton in Versova and Mahim Creeks

Versova		Mahim	
1982	1996	1982	1996
Nitzschia	*Oscillatoria*	*Oscillatoria*	*Skeletonema*
Thalassiosira	*Skeletonema*	*Nitzschia*	*Nitzschia*
Oscillatoria	*Nitzschia*	*Thalassiosira*	
Chaetoceros		*Navicula*	
Navicula			

Zooplankton

The creek systems are characterised by wide fluctuations in the standing stock of zooplankton (Figure 10.3) with lower faunal diversity as compared to openshore. Unlike polluted creeks, openshore water reveals higher incidence of siphonophores, lamellibranchs, appendicularians and fish larvae (Nair and Govindan, 1986). Abundance of carnivores like medusae, ctenophores, mysids and polychaetes are the characteristic of polluted creeks. A reduction in zooplankton standing stock of Versova and Thane creeks during 1995-96 as compared to 1982-85 is evident from Figure 10.3. A marginal increase in zooplankton biomass at Mahim during 1995-96 (DOD, 1996) than 1982-85 could be due to the dominance of carnivores like medusea and ctenophores which are abundant in the polluted creek systems (Nair and Govindan, 1986).

Macrobenthos

The standing stock of macrobenthos in terms of population reveals wide fluctuations in the polluted creek systems around Mumbai. The macrobenthic biomass however, is relatively higher in Mahim and Thane creeks as compared to Versova creek (Figure

10.3). Benthic biomass is moderate in the openshore regions though the faunal group diversity shows improvement. The high standing stock of benthos usually encountered in the polluted creeks could be due to the enrichment of sediments with high detrital materials through high primary productivity of the water column, and the particulate matter rich in organic matter associated with domestic wastewater settling on the bed. The benthic filter feeders and deposit feeders conveniently make use of the rich carbon availability through the food chain. Hence, in many instances (Varshsney, 1982) the enhancement of primary production subsequently increases the benthic faunal potential. However, inspite of enhanced benthic faunal potential of selected pollution tolerant groups, a decrease in species diversity (H) of polychaetes at Thane (0.67) as compared to coastal waters (1.84) away from the creek, is evident (Annie and Govindan, 1995). Species composition of polychaetes in general is inhibited at the zones which are under severe pollution stress which can be attributed to the signs of ecological imbalance of the ecosystem (Armstrong *et al* 1981).

Pollution leads to a lower growth rate condition index and percentage edibility of intertidal molluscs from Mumbai (Krishnakumari *et al.,*1990; Fernandes 1991). Higher levels of bioaccumulation of trace metals and PHC in the tissues of intertidal gastropods from highly polluted regions like Worli as compared to Madh and Colaba have also been reported (Fernandes 1991).

Fishery

The openshore waters (DOD, 1992) sustain moderate fishery potential (5.25 kg/h). The abundance of fish eggs and larvae increases towards the offshore from the coastal inshore and creek regions. Higher productivity at the primary and secondary levels is expected to increase the pelagic fishery potential especially within the creek systems. However, the high productivity at the primary and secondary levels in the creek systems is mainly induced by pollution and supported by proliferation of pollution tolerant species which need not necessarily end up with high teritary production. In other words, the high productivity in lower trophic levels due to pollution is not entirely utilized at tertiary level to induce the expected fishery potential. Since most of the tertiary producers are highly mobile and sensitive to environmental stress, they can easily avoid such polluted systems which cause under utilization of prevailing high biomass at the lower trophic level resulting in low ecological efficiency (Varshney *et al.* 1982-83) interms of transfer co-efficient values (Table 10.6). Hence, these polluted creek systems with poor fishery potential should be classified as special ecosystem where the normal trophic conversion factor of 10% is not applicable.

Table 10.6: Observed and Estimated Biological Productivity in Different Creeks.

Station	Observed PP (tc/km^2/y)	Estimated SP (tc/km^2/y)	Observed SP (tc/km^2/y)	TC (%)	Estimated fishery potential (tc/km^2/y)		
					Based on PP	Based on SP	Mean PP& SP
Versova creek	172.6	17.3	7.9	4.6	12.9	5.9	9.4
Versova offshore	138.9	13.9	10.6	7.6	10.4	8.0	9.2
Mahim creek	120.9	12.1	7.2	5.9	9.0	5.4	7.2
Thane creek	194.3	19.4	17.2	8.9	14.5	12.8	13.7

Note: PP- Primary Productivity; SP- Secondary Productivity; TC- Transfer Coefficient

Bioaccumulation of trace metals

Marine organisms at different trophic level indicate bioaccumulation of selected trace metals like Cr, Mn, Fe, Co, Ni, Cu, Zn, Cd, Pb and Hg through the marine food chain. Zooplankton like copepods and chaetognaths from various creek systems around Mumbai shows (Nair and Govindan, 1986) different levels of body burden of trace metals (Figure 10.4). Higher body burden of metals like Cu and Mn at Mahim, and Co and Ni at Thane is evident. Copepods concentrate higher levels of many of these trace metals as compared to chaetognaths indicating regional and species differences in bioconcentration. In case of benthic invertebrates, the concentration of metals is relatively more in polychaetes than decapods (Figure 10.5).

Bioaccumulation of Fe, Zn and Pb in polychaetes is more in Thane creek as compared to Mahim and Versova creeks. The decapods concentrate higher levels of Cd and Pb in Thane creek as compared to other creek systems. The bioaccumulation of metals at the higher level of food chain, viz. the fishes, shrimps (crabs and prawns) and molluscs in Thane creek also occur. Bioconcentration of metals in molluscs is relatively more as compared to shrimps and fishes. In general, the concentration of various trace metals in fishes and shrimps is considerably low as compared to marine organisms at the lower trophic levels like zooplankton and benthos. However, bioconcentration of metals like Cu, Zn, Cd, Pb and Hg in fishes from Thane creek, though low, pollution induced metal burden at the tertiary level of the food chain is quite clear. The increase in pollution load is likely to cause increase in body burden of metals in marine organisms of the creek system which is not desirable. Proper pollution control and monitoring of various creek systems around Mumbai are become imminent.

CONCLUSIONS

Domestic and industrial wastewater released indiscriminately have severely deteriorated the water quality of Mahim, Versova and inner Thane creeks and Ulhas estuary. Intermittent or total depletion of DO, active denitrification and liberation of hydrogen sulphide in onshore waters indicate the severity of pollution due to the fluxes of organic loads which far exceed the assimilation capacity of these creeks. The openshore waters though sustain the signatures of contaminants drained through the creeks, bays and estuaries are relatively clean.

Transport of pollutants from the water column to the bed has modified the sediment character to a marked extent in the inshore waters with high levels of trace metal, OC and TP and sometimes PHC in the surficial sediment.

The biotic features indicate very high bacterial contamination, enhanced concentration of phaeopigments and phytoplankton counts with low generic diversity, wide fluctuations in zooplankton standing stock with higher abundance of carnivores and low faunal diversity, increased abundance of macrobenthos with reduced diversity and lower fishery potential especially of the polluted creek as compared to offshore coastal system. Bioaccumulation of selected trace metals in zooplankton, benthos and fishes especially in the creek systems reveals slow build up of metal burden through the food chain. Hence, alternate mode of waste disposals with strict pollution control measures to ensure the health of the creek systems around Mumbai is warranted.

ACKNOWLEDGEMENT

The authors are thankful to Dr. E. Desa, Director, NIO for facilities and encouragement.

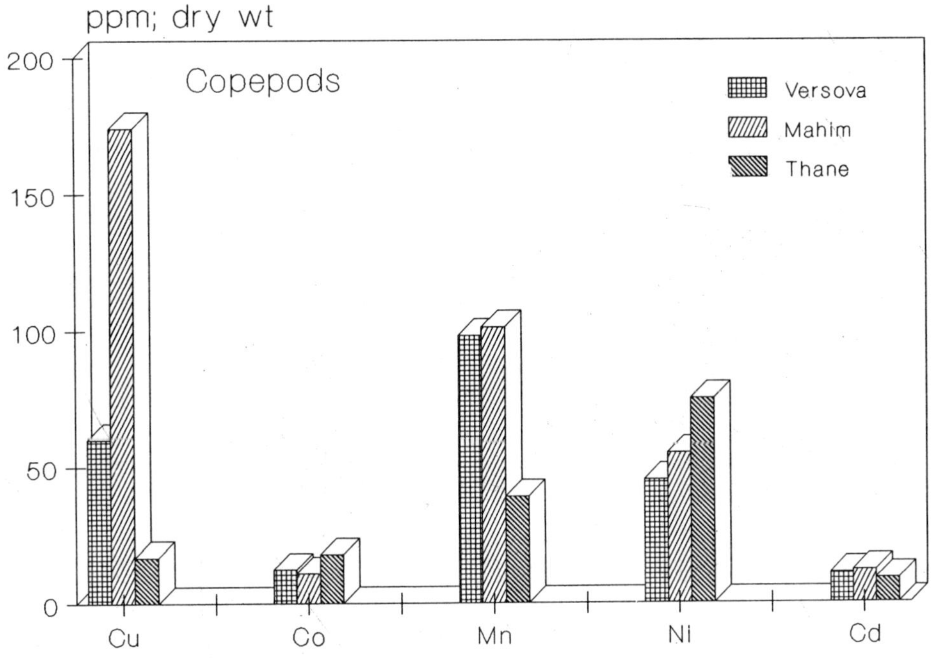

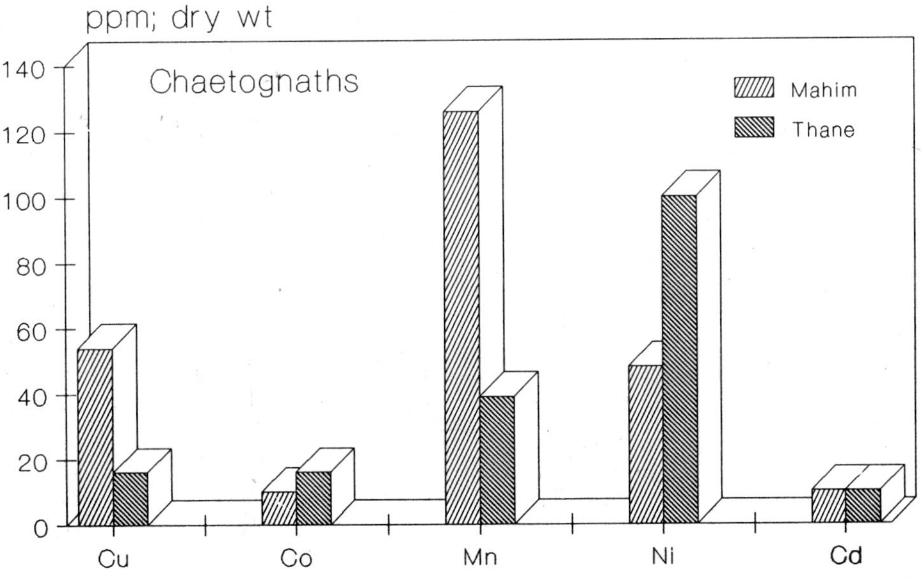

Figure 10.4: Average Concentration of Trade Metals in Zooplankton of the Creek Systems.

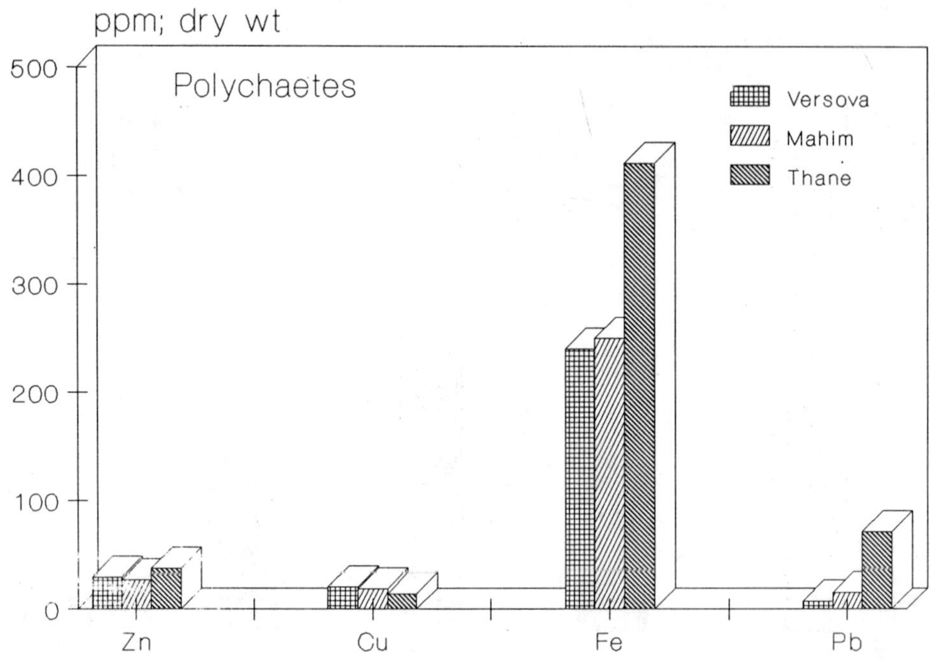

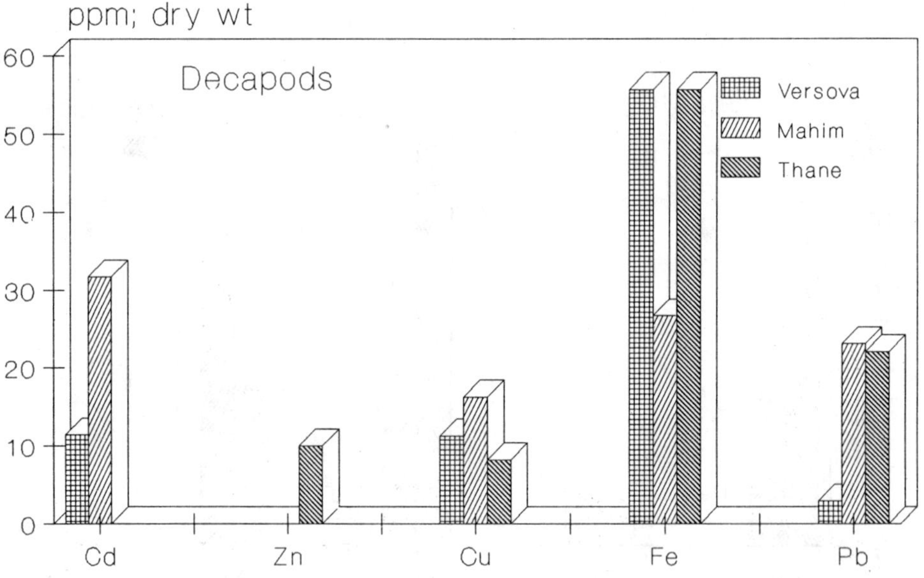

Figure 10.5: Average Bioaccumulation of Trace Metals in Benthic Organisms of Mumbai (Varshney 1982).

REFERENCES

Annie Mathew and Govindan, K. (1995) Macrobenthos in the nearshore coastal system of Bombay. Proceeding of the National Academy of Science, India, **65 (B), IV**: 411-430.

Armstrong, J.W., Jhom, R.M and Chow, K.K. (1981) Impact of a combined sewer overflow on the abundance, distribution and community structure of subtidal benthos. *Marine Environmental Rsearch* **4**: 3-23.

Bhosale V. and Sahu, K.C. (1991). Heavy metal pollution around the island city of Bombay, India. Part II. Distribution of heavy metals between water, suspended particles and sediments in a polluted aquatic regions. *Chemical Geology* **90** : 285-305.

DOD 1992. Coastal Ocean Monitoring and Prediction System (Technical Report). Department of Ocean Developement, New Delhi.

DOD 1996. Coastal Ocean Monitoring and Prediction System (Technical Report). Department of Ocean Developement, New Delhi.

Fernandes, J.L. (1991). Impact of marine pollution on selected intertidal gastropods of Bombay. Ph.D. Thesis, University of Mumbai, 1-27.

Krishnakumari, L., Vijayalakshmi R. Nair and Govindan, K. (1990). Some aspect of growth and condition index of *Saccostrea cucullata* (Born), *Cerithium rubus* (Desh) and *Tellina angulata* from Bombauy coast. *Journal of the Indian Fisheries Association,* **20,** 21-29.

Ram, J.J. (1985). Studies of phytoplankton in polluted and unpolluted aquatic environments of the north west coast of India. Ph.D. Thesis, University of Mumbai, 1-519.

Sabnis, M.M (1984). Studies of some major and minor elements in the polluted Mahim river estuary. Ph.D. Thesis, University of Mumbai, 1-288.

Sahu, K.C. and Bhosale, V. (1991) Heavy metal pollution around the island city of Bombay, India, Part I: Quantification of heavy metal pollution and the recognition of environmental discrimitants. *Chemical Geology* **90** : 263-283.

Sharma, P. Borole D.V. Zingde M.D. (1994). ^{210}Pb based trace element fluxes in nearshore and estuarine sediments off Bombay, India. *Marine Chemistry,* **47**: 227-241.

Srinivasan, M. (1990), Heavy metal pollution along the Thana creak and Ulhas estuary around Bombay Island. Ph.D Thesis, Indian Institute of Technology, Mumbai, 1-188.

Varshney, P.K. (1982). Biological productivity of polluted and unpolluted environments of Bombay with special reference to benthos. Ph.D. Thesis, University of Mumbai, 1-166.

Varshney, P.K., Govindan, K., Vijayalakshmi R. Nair and Desai, B.N. (1982-83). Biological productivity and fishery potential in the coastal waters of Bombay. *Journal of the Indian Fisheries Association*, **12** & **13**, 35-43.

Vijayalakshmi R. Nair and Govindan, K. (1986). Biological characteristics as a part of pollution monitoring studies. Proceeding of the National Seminar on Mussel Watch (1986), University of Cochin, 86-90.

Zingde, M.D. (1989). Environmental status of the coastal marine environment of India. In : Management of Aquatic Ecosystems (Eds. V.P. Agarwal, V.P. Desai and S.A.H. Abidi), Society of Biosciences, Muzaffarnagar, India, 37-57.

Zingde, M.D and Desai, B.N. 1981. Mercury in Thane creek-Bombay harbour. *Marine Pollution Bulletin* **12,** 237-241.

Zingde, M.D. and Sabnis, M.M. (1994) Pollution induced tidal variability in Water Quality of Mahim estuary. In : Environment and Applied Biology (Eds. V.P. Agarwal, S.A.H. Abidi and M.D. Zingde). Society of Biosciences, Muzaffarnagar, India, 277-298.

Zingde, M.D., Bhosale, N.B., Narvekar, P.V. and Desai, B.N. 1989. Hydrography and water quality of Bombay Harbour. In : Environmental Strategies and Biosciences. (Ed. R. Prakash), Society of Biosciences, Muzaffarnagar, India, 37-58.

11

INTEGRATED COASTAL ZONE MANAGEMENT IN GUJARAT

Ashwani Kumar

ABSTRACT

Human beings always try to attain the status of food security which is reflected into intense efforts of cultivating the soil up to the level that is endangering the balance of resources. But in the last few decades, it has been felt that the land resources can not serve the purpose of food security for longer periods under the 'business as usual scenario.' Also the exploitation of land, which is already degraded and over-burdened can not be continued further. The problem of population increase and scarcity of resources has become more serious in developing nations. To meet the future requirements, renewable type marine resources can be the focus of attention. These resources are available all along the coastline of India. Integrated management of these resources can solve several problems in the country. This chapter explains the environmental issues of integrated coastal zone management with the help of a case study in the state of Gujarat.

INTRODUCTION

The unique characteristics of the coastal zones is the interface of two resources land and water that make it suitable for most of the human needs. The nations possessing greater technical and economic capabilities have already started exploiting their own and also of the under developed neighbour's coastal resources indiscriminately. This resulted in conflict of interest among different countries. International treaties finally established two hundred miles distance from the shore as exclusive economic zone (EEZ) which greatly augmented the quantities of living and non-living marine resources within coastal nation's jurisdiction and decreased access of developed nations into the EEZ of developing nations.

Coastal zones have been put under various uses through various developmental activities like industrial development, exploration, navigation, export centres, fisheries as well as human settlements. The coastal areas of India having high diversity of flora and fauna are in the midst of economic development. These areas of country are being over-utilised in many ways affecting the renewable capacity of resources as exploitation rate has exceeded the regeneration rate. With this hypothesis, the present study attempts to identify the major environmental issues of coastal area within the framework of the approach of Integrated Coastal Zone Management (ICZM). In general, the management of coastal zones ideally involves:

- ensuring the well being of coastal people,
- protection and conservation of the environment,
- optimisation of coastal resource exploitation within the homeostatic limits,
- rationalisation of various activities, and
- Overall improvement of ecosystems.

The Coastal Zone Management is relatively new to India, and little work has been done in this multi-disciplinary field because of prevailing sectoral approach of planners policy makers and private agencies till now. Thus, to cater to the above requirement, the present study aimed at addressing the environmental issues through ICZM for ecosystem conservation with the following objectives:

- To assess the environmental status of area.
- To identify/establish various linkages between human activities and environment.
- To investigate the emerging environmental issues.
- To investigate the developmental proposals and their impact on the area and the present environmental issues.
- To suggest possible tools to handle these issues through ICZM approach.

METHODOLOGY

The search for a proper methodology became the first important critical issue at an early stage for increasing the level of understanding its diversified nature. The objectives of such a highly multi-disciplinary nature of methodology within the constraints of lack of data, information of scientific nature at micro level, inefficient and poor documentation conditions due to its own reasons etc. compelled to adopt an alternate way to derive inferences, such as the feedback from the population, to understand the actual ground realities and to capture the undocumented trends. The theoretical methodology developed based on the literature survey highlighting the few key issues in the Coastal Zone Management,

- The greatest challenge in ICZM is to define both ecologically sound and politically manageable unit.
- Any eco-development, management or conservation programme cannot be delinked from the history and evolution of the ecosystem and physical features of the study area.
- The consideration of human demand of coastal resources is not separate from ICZM but in fact it provides a mechanism to check the over exploitation.
- Consideration of regeneration capacity is a major determinant of the level of exploitation
- Because the implementation of any resource management requires the public support with varying degree, it calls for developing and understanding of social environment, traditions, culture and practices of people and identifies the beneficial and harmful issues.

The attempt to translate the first key point for the coastal zones, the regulatory definition of five hundred metres CRZ was found ineffective. This is because either potential impacts of sources stay outside of the zone or stay within it for a short duration. In order to delineate the coastal area capabilities of human beings traveling on feet was taken as one of the prime criteria. It was found that most of the fishermen who are one of the major actors of coastal areas travel on feet a distance of one to seven km daily to carry out the activities. The coastal areas were taken as settlement all along the coast or places where tidal effect is felt through creeks.

At the next stage, a group of 25 villages on the basis of the above criteria was selected for analysis to demonstrate the disparities between coastal areas and main land in terms of different Socio-economic amenities and parameters. For this purpose secondary data mainly census report were used. As the second step, different kinds of activities were identified and concentrated upon based on their importance for the impacts on coastal ecosystem and resources. Such activities identified in the study area were salt industries, fishing, newly conceptualised aquaculture, afforestation etc. The ecosystems blessing the area were mangroves and estuaries.

DETAILS OF THE STUDY AREA

The study area focuses on Jambusar taluka of Bharuch district located at the Gulf of Khambat in the state of Gujarat. The coastline of Gujarat is about 1600 Kms, which is about 21.3 percent of India's coastline. Gujarat has comparatively the longest coast line among all the States of India but its natural potential was not put to the full advantage and age old major ports like Lothal, Mandvi, Bharuch, Surat, Ghogha, Dahej, etc. were not taken into serious consideration. The study area selected was the coastal area of Jambusar taluka of district Bahruch in Gujarat having the following profiles:

Socioeconomic

On the demographic front, the population of Jambusar taluka increased from 1,58,316 in 1981 to 1,64,262 in 1991 with decadal growth rate of 3.7 percent. The drastic decrease in growth rate in the taluka as compared to 1971-81 indicate the possibility of out migration due to the prevailing economic poverty in the area. Calculations done on coastal villages show literacy rate of 54.6 percent which is marginally lower than talukas' average literacy rate 55.14 percent in the year 1991.

The analysis done on the coastal belt revealed

i. higher barren land in coastal belt (49.3 percent) as compared to taluka average (29.6 percent) 1993-94),

ii. lower cropped area of 42.8 percent while that of taluka stands at 60.87 percent for 1993-94,

iii. non-agricultural land 3.6 percent lower than taluka average due to lower level of development of villages along coast and

iv. fodder land marginally lower at 2.3 percent than taluka average of 2.45 percent.

Occupational Structure

The population involvement in agriculture sector is facing a decline both for cultivators and agriculture labourers (from 38% to 32% and from 44.7% to 36.5%) during 1981 to 1991, respectively). The reason for the above unusual trend can be chiefly attributed to the shifting of population from agriculture based to industry based occupations. Decrease in labour force indicate the improvement in employment conditions. Maximum demand of non-agriculture based occupation is in coastal areas due to under developed agricultural conditions.

Based upon the preliminary study in the area, following linkages were identified for detailed analysis. The parameters selected are those, which play critical linkages in human-environment relationship and are primarily classified into three major categories.

i) Land resource based linkages

-Agriculture

-Salinity

-Erosion deposition

ii) Water resource based

- Irrigation

- Drinking Water and Domestic Water

- Fisheries; Marine Inland and Aquaculture

- Coastal Water Pollution

iii) Vegetation & Ecological Importance

- Mangrove forests
- Flora of area; Forests and crops
- Livestock and cattle pressure

FINDINGS AND MITIGATIVE MEASURES

Based upon field visits, discussion with concerned communities, voluntary agencies and experts in the field, the following points have emerged. Findings for all important issues and appropriate policy measures are beng outlined.

Agriculture Sector

1. Agriculture sector is under constraints of lack of inputs, water and fertilizers resulting in lower yield and hence lower economic gains
2. Technical innovation is not experienced by the farmers of the taluka and low level of mechanisation exists resulting in lower yield.
3. Innovation in new agricultural methods is virtually absent in spite of the efforts of the NGOs working in the Taluka.
4. Agricultural pollution is absent at present. However, the possibility of a higher level of fertilizer and pesticide use exists.
5. Narmada dam will certainly give a remarkable relief to the problem of water scarcity for irrigation.
6. Coastal areas, specially northern fringe, is under high livestock pressure with lower fodder area
7. Coastal Zone is much more vulnerable to food security than main land.

Suggested Measures

- Water scarcity can be solved by building the small water storage facilities in low depression areas through watershed approach.
- Agricultural inputs should be increased to yield higher output.

- Alternate to the agricultural practices like silviculture, should be implemented to increase self-reliance.
- New salt-tolerant species of crop should he adopted by germination techniques and bioengineering.

About the Salinity

i. Jambusar lies in the region of inherent salinity whose origin is marine bed soil coupled with salinity ingress of seawater.

ii. Agricultural fields are being increasingly affected by the salinity, which has engulfed the lands not only along the coast but also mainland side.

iii. Partial effect of salinity has started to mark its presence in Vadodara district also.

iv. Human made salinity is not seen because of lack of irrigation facilities and lack of good ground water potential. However, with the Narmada project this problem is expected to increase in the view of poor drainage condition and existence of salt horizon within 1.5 m depth.

Suggested Measures

- Improved Management of Salt-affected Lands

Leaching of the excess salts which has been a success elsewhere in the country is not feasible here because of unavailability of even moderately saline water in sufficient quality. Thus, the only solution, which emerges, is rainwater utilization for these purposes and watershed approach to increase the water recharge.

- Irrigation

In saline soil irrigation cycle is changed from an 'extraction dominated' process to an 'infiltration irrigation' for full yield in alkali/saline soils, a high frequency application and low magnitude irrigation practice is required.

- Drainage and Leaching

Effective leaching system can be used only with proper drainage. In case of financial constraints, restoring adequate levelling and bunding can contribute to 50 to 70 percent rainfall towards leaching and 90 percent of the salts can be leached below the depth of 45cm in the soil.

- Agronomic measures

The practice, which assists in maintenance of soil productivity and helps sustaining the yield stability of the crop on the soil, is called agronomic practice. The crop, their varieties, the cropping sequence are to be decided according to their tolerance to salinity for the higher economic gain.

- Other practices

Pre-sowing irrigation leaches the root zone soil and improves the germination and early growth when plants are not very tolerant to salinity. But in the study area it can not be practiced due to non-availability of superior quality of water that is most essential for the purpose. Increased seed rate increases the grain yield by 18 percent under saline and higher water table conditions. This can be attempted in the southern part of taluka where water table is within 3 to 5 m depth.

The crop plantation on ridges oriented in the NE and SW directions should be tested on the saline soil. Where desalination takes place in NW slope as the result of salt movement to SE slope, which receives greater intensity and direction to solar radiation, ridge constructed should be of 25 cm high with base angle of 60 degrees. In the coastal fringes, the kharland should be taken for trench plantation.

About Erosion

i. Lack of the vegetative cover for major part of the year in the taluka is the main contributor to the erosion hazards.

ii. Jambusar taluka has an erosion rate of 0.2-0.5 kg/m^2/annum which is not very high, but being classified as a sensitive area efforts should be made to keep the rate within the present limit, if not reduce it.

iii. The erosion taking place along the Mahi river is resulting in a cliff off the Mahi bank.

Suggested Measures

- For controlling soil erosion in the area of study, proper agronomic and soil management measures should be given a high priority.
- Measures should cover protection against direct raindrop impact.

- Increase in soil infiltration capacity and roughness to reduce runoff and stability. This can be attained by increasing the vegetative cover along the coast, creeks and in the wastelands.
- Mechanical measures can not control detachment of soil particles and only the transport of soil particles can be reduced by flow of any excess water and wind that arises but mechanical measures are expensive and skill intensive.
- The Mahi banks should be given a high priority for afforestation to control the erosion.

About Marine Fisheries

i. Marine fishery catch has reduced over the last few decades which can be attributed to a series of reasons such as increasing silt load in coastal water, decrease in sea depth, deteriorating conditions of the fishermen and mangrove loss.

ii. Marine fishermen are shifting to other industries from the fishing activities.

ii. None of the institutions/agencies is found active with the concentrated efforts for fisheries and fishermen population in the taluka. Only a few fishermen co-operatives exist but they are ineffective due to lack of guidance.

iv. Lack of credit and marketing facilities.

Suggested Measures

- Institutional arrangements such as constituting taluka level Agency/Co-operative to look into various aspects of fisheries and problems of fishermen population.
- Credit facilities, which can arrange finance in the form of subsidy from the Government of Gujarat to basically, finance the village pond and other tank fish culture should be introduced. The arrangement to encourage the marine fishermen and separate funding scheme should be worked out to link with better marketing facilities.

About Inland fisheries

i. Only the bigger ponds of more than 5 ha area found to be most beneficial.

ii. Lack of co-operatives for fisheries.

iii. Lease arrangement on short term basis makes this sector uninteresting for people and banks.

Suggested Measures

- Water availability in pond depends on soil characteristics like in Magnad; sandy clay dominates result into faster depletion of pond water. Such ponds should be used after brick lining only.
- Inland pond culture requires 8 months to attain fetchable size. But most of the ponds in the taluka (62 percent) are of short term water availability All ponds of perennial nature should be used for prawn culture of short duration for semi-extensive fisheries.
- Formation of more fishermen co-operatives should be encouraged and marine fishermen should be given preference along with the normal preferential group of backward classes to utilize their skills and potential for income increasing.
- As some of the ponds require higher capital investment with some gestation period. So such pond should be identified and a longer lease arrangement at the first negotiation itself should be incorporated.

About Aquaculture

i. Aquaculture is coming up in the area following the eastern coasts, and will prove to be an economic boon for the area with a high internal rate of return.

ii. Marketing and seeding facilities are not adequate.

ii. The quality of the effluent is acceptable and not violating the standards at present but may become a critical issue in the long run and which may need proper monitoring.

iv. Health hazards are presently not there but future implications are important.

Suggested Measures

i) Buffer Zone

A study by the National Environmental Engineering Research Institute recommended the setting up of a green buffer zone of over 250 m between the prawn farms and other land use. The plan is to construct an earthen bund around prawn farms, with a drainage channel on the farm side to allow the seepage of pond effluents into the channel leading it to the sea or estuary. The bund should be planted with 5-6 rows of casurina trees or Prosopis juliflora to absorb saline water and should have a clay lining to prevent seepage. But because at present the corning up farms are not intensive in nature and the area under such is also not large, so the buffer zone requirement is not proposed in the area. However future strategies should include the buffer zone concept to achieve a better environmentally managed aquaculture in the area.

ii) Eco-restoration Fund

The state government should propose compensatory eco-restoration fund in which each industrial unit as well as aquaculture farms would deposit money, which can be utilised for the plantation, watersheds and other measures. Mangrove plantation programme can seek help under this scheme.

About Industrial Pollution

i. Mahi and Dhadhar rivers are under a risk of getting pollution more than that can be tolerated by marine life.

ii. The effluent from the channel is lifted for irrigation at a number of places and can lead to a higher conc. of heavy metals in the food grains.

Suggested Measures

- The surveillance against the possible interception should be strictly implemented.
- The farmers should be made aware of the ill-effects of the toxic metals present in the effluents.
- At least the trace metal concentration in the effluent should be reduced to make it harmless for using it for irrigation.

About Mangroves

i. Some conflicts exist in the classification scheme adopted by different agencies giving different coverage and pose a difficulty in management of mangroves.

ii. In spite of defined conflicts, mangroves are found to be depleting at the rate of 30 percent per decade.

iii. The camel grazing is the only utilisation pressure acting on the mangroves of Jambusar.

iv. The Khambhat oil potential if explored can lead to mangrove loss through higher oil and grease discharge during transportation and refining operation.

v. People are unaware of the ecological importance of mangrove but they are against the camel grazing and have demanded the protection of this forest.

Suggested Measures

- The camel grazing practice should be immediately banned and high penalty should be imposed on violation of it
- Public awareness regarding the role of mangroves in fisheries boosting and erosion reduction should be increased.
- A surveillance cell of forest staff and the local people should be formed and should be equipped with a small temporary staying facility at the nearest possible point from the mangrove (such as at Devjagan temple).
- The plantation drive should be organized with the people to increase the mangrove coverage under the Forestry schemes of Government.
- Higher wages are required to motivate the people.
- Formal education through field visits and discussion should be given to the plantation teams.
- Species present in the area should be used preferably to match the local condition.

CONCLUSIONS

Manangement of coastal zones is a crucial issue in the state of Gujarat. The case study revealed termendous difference between the environmental and socio-economic trends in the costal areas and the main land. Problem of increased pollution due to agriculture and industrial activities is alarming. In addition, over exploitation of fisheries and other coastal resources are of serious concern. Integrated coastal zone managemnt requires joint efforts of the government and non-government organisations and local people.

12

HUMAN INTERFERENCE ALONG THE COAST OF GOA

Antonio Mascarenhas

ABSTRACT

Till the 1970's the goan coast was largely pristine and unsullied, as most coastal activities were located away from and behind sand dunes. A distinct harmony prevailed between man and Nature. During the last two decades or so, the promotion of tourism, increase in population and construction activities have created a heavy demand for resources resulting in congestion on coasts, increase in density of constructions and pressures on infrastructure. We note large scale conversions of dune belts leading to indelible alterations of coastal landscape, in general, and dune ecosystems, in particular. Lack of implementation and enforcement of legislations has led to blatant violations of CRZ laws and uncontrolled growth along certain sectors of the coast. Major impacts are seen in the form of damage to dune vegetation, permanent elimination of coastal sand dunes and (localized) beach erosion. Major impacting factors are constructions on dune belts, mining of dune sands and roads along sandy areas. Excessive pumping of ground water has induced salt water ingress, although seasonal in nature. Other human impacts such as beach shacks and recreation appear to be less harmful. Building activity has now gradually invaded coastal hill slopes some of which are already degraded. Maintaining of appropriate free spaces and compulsory set back lines, together with a stringent enforcement of CRZ legislation are imperative so as to preserve the remaining shreds of the coastal environment.

INTRODUCTION

The coastal zone of Goa had been exclusively used for agriculture, farming, shell fishing, traditional fishing and low key recreation (Dume, 1986). Goans of yesteryears used the shoreline for fishing by using hand-cast and hand-pulled nets; traditional fishing was the main economic activity of coastal populations. They never occupied dunes or beaches. The only identifiable structures along the shore were a few cabins and thatched huts that housed sea going canoes, some of which can still be seen today. The large plain areas behind the dune belts were used for farming and paddy cultivation, activities which are common at certain places even at present. Recreation was restricted to a few beaches only.

Although little was scientifically known those days, our forefathers had somehow grasped the sanctity of Nature, the sand dune ecology for example (Lobo, 1988). Till the 1970's, there was no evidence wherein sand dunes were razed to build houses. Ancestral mansions, roads, traditional villages, and any major activity such as paddy cultivation were all located at safe places away from and behind the dunes. Although obscured by modern development, this type of age old planning can still be identified. Such places landward of dune belts thus remained as areas of peace and tranquillity. The coastal zone environment was largely pristine, well preserved. Therefore, such a setup could be cited as a unique example of the harmony that prevailed between man and Nature (Lobo, 1988; Mascarenhas, 1990; Alvares, 1993; Mascarenhas et al., 1997).

Human pressures on the beach - dune systems started during the 1970's when tourism was declared a potential revenue earner for Goa, a place that was projected as a major tourist destination. During 1971 - 1991 period, almost 80% of the urban growth was, and still is, concentrated in the coastal talukas (D'Souza et al., 1988; Wilson, 1997). Moreover, during the last decade, a large number of people chose to move towards the coast as can now be observed at Caranzalem or Dona Paula, for example. Since then, there has been a dramatic growth and a proliferation of innumerable coastal resorts, residential flats, dwellings, small restaurants, roads and beach shacks along and across the sand dune belts. Several coastal areas have thus changed from virtual wilderness in 1970's to haphazardly developed stretches, full of concrete buildings and related structures, in less than 20 years (Mascarenhas, 1997; Sawkar et al., 1998). Several shore fronts have been built in such a manner that they bear little resemblance to the coast that formerly existed. All these activities resulted in haphazard growth and overcrowding of coastal regions that we see today.

Therefore, the advent of tourism and population increase, coupled with building activity and modern societal demands, have resulted in large scale changes in the geological and ecological setup and indelibly altered ecosystems, land use patterns and the coastal zone landscape (Lobo, 1988; Alvares, 1993; Mascarenhas, 1997; Mascarenhas et al., 1997; Narayan, 1997; Sawkar et al., 1998; Mascarenhas and Sawkar, 1998; Rosario, 1998). Considering that the open sea front of Goa is being built up at a very fast pace, it is not known in what way the heavy developmental activities are affecting the sandy areas, in general, and the dunes, in particular. Therefore, this study is aimed at understanding the influence of man along the coastal strip including areas around estuaries. The purpose of this chapter is to identify issues of conflict, highlight various anthropogenic impacts during the last two decades and delineate areas of concern along the open coastline of Goa.

COASTAL ECOSYSTEMS OF GOA

The open sea front of Goa is characterized by a combination of beaches, rocky shores and headlands (wooded or bare) which protrude into the sea. These promontories are composed of basaltic or metamorphic formations capped by laterite and intruded by basic igneous dykes. Linear sandy beaches are located between promontories; sandy pockets are often found at the base of coastal hill slopes. The width of the beaches varies between 50 and 180 meters. Of the 105 km long coast, more than 70 km comprise sandy beaches, all backed by several rows of 1 to 10 m high sand dunes which extend almost half a kilometer or more before merging with the hinterland coastal plain.

Five key coastal stretches are characterized by conspicuous sand dune complexes as follows (Mascarenhas, 1997).

1. Querim - Morjim sector with pristine beaches and turtle nesting sites (Anonymous, 1997a), including three creeks lined by mangroves and located behind sand dunes;
2. Chapora - Sinquerim belt;
3. Caranzalem - Miramar (Mandovi estuary), the most prominent dune belt within the estuaries of Goa;
4. Velsao - Mobor linear stretch being the longest strip of the most exquisite dune system of the entire coastal zone of Goa, and

5. Talpona - Galgibaga strip, presently a pristine area. In addition, the coastal stretches also consist of several sandy areas and secluded coves backed by cliffs, rocky shores, headlands or promontories and wooded or bare hill slopes. Some uninhabited islands with an appreciable forest cover are found off Goa.

From north to south, the coastal zone of Goa is traversed by seven major dynamic estuarine rivers and four minor river systems. The Combarjua canal is the only connection between two such major rivers. Most of the major rivers which cut across hinterland formations originate in the western Ghats, across the border (Alvares, 1993). All the rivers are lined by dense mangroves, except at places. Several islands and shoals inhabited by thick mangroves are found within the rivers, the Chorao island being the most prolific. The tidal effect can be felt more than 40 km in the hinterland.

Throughout the course of these rivers, one can find an intricate network of creeks and backwaters scattered all over the coastal zone. A luxuriant growth of mangroves (some of which are degraded) and associated swamps can be observed in most water bodies. The most prominent and extensive backwaters with mangroves are located east of the capital city of Panjim.

The coastal plain of Goa varies in width. In the northern part, it extends to about 35 km, virtually to the base of the western Ghats, whereas it is about 20 km wide in the central part. In the southern part, the coastal plain is less prominent, identified only adjacent to the two rivers; this sector is mostly occupied by evergreen forests.

In general, the coastal plains comprise an intricate system of wetlands, tidal marshes and cultivated paddy fields, all intersected by canals inland lakes, bays, lagoons and creeks. All the rivers and the extensive backwaters in the hinterland are governed by regular tides, which raise or lower water levels by 2 or 3 meters daily. The prominent lowlands found adjacent to most of the rivers are locally known as "khazan lands", a term which denotes land reclaimed by gradual filling of the shallow seas. Most of these lowlands are almost at and even below sea level.

METHODOLOGY

Issues discussed in this chapter form a part of a larger study being conducted over the past two years; this work is still continuing. Extensive field checks have been regularly carried out along the entire coastal zone of Goa; more than 200 sites (including backwaters) were inspected and surveyed, and detailed physical data and ground truth observations have been collected with which the areas of concern can be identified. Various types of geological samples of different landforms including dune sands were also collected. The photography of natural features of various coastal ecosystems has been undertaken to justify their ecological importance and their intrinsic value, and also to depict the present coastal scenarios. Violations of coastal zone environmental regulations have been noted where they exist; this exercise mostly consists of identifying man-made structures in wrong places along the coastal tracts. Existing toposheets and recent satellite imageries were also referred to; these specific details are being published elsewhere.

Various official and other reports, published and unpublished, have been made use of for comparison with our data. Discussions with respective officials, planners, bureaucrats, scientists, politicians, senior citizens and natives or locals concerned with coastal issues are of immense use from the point of view of problems and conflicts, planning and management and also in reconstructing past scenarios for a comparison with the changes observed at present.

PROBLEMS AND CONFLICTS

Since 1982, there have been a number of official reports on various aspects of coastal issues, policy matters regarding coastal activities, land use and tourism. The Master Plan for Tourism development (Anonymous, 1987) describes infrastructural problems and classifies the development into tourism development areas, tourism resort areas and village tourism areas. The Regional Plan for Goa (D'Souza et al., 1988), a more comprehensive document, highlights detailed town and country planning policies, aims at rationalisation of disparities in development, discusses integrated rural development, environmental conservation and measures for the improvement in the quality of life.

Although these documents were published in public interest, they are not readily available for public reference, and unfortunately, all environmental plans have been shelved. All these documents have therefore turned into an exercise in futility (Anonymous, 1998b). Moreover, the government brought an amendment empowering itself to modify the Regional Plan whenever required; this plan is now obsolete. Thus, a sincere attempt to advocate sustainable development has been ruined.

In Goa, human pressures on the coasts are being felt since the early 1970's when the revenue potential of tourism was identified. The first starred beached resort was set up at Sinquerim in 1974, the second at Dona Paula in 1982 and another in Majorda in 1984. Since then there has been a dramatic growth and a proliferation of innumerable edifices such as coastal hotels and resorts, residential flats, dwellings, small restaurants, beachside bars, roads and beach shacks along and across the coastal zone, but most of which are located on sandy areas as dunes. Some of the surveys indicate that the growth is uncontrolled (Sawkar at al., 1998; Chari, 1998; Mascarenhas and Sawkar, 1998) and as such several coastal areas are overcrowded (Wilson, 1997). This growth resulted in an undesirable over urbanization of coastal regions with an associated loss of biodiversity and deterioration in the quality of life.

Since there is a heavy competition over limited coastal space and resources with a consequent rise of land prices, the coastal space utilisation has now turned into a delicate issue mostly because of various types of conflicts on different topics (Alvares, 1993; Anonymous, 1995). Discordant views are being regularly expressed between real estate developers, builders, planners, policy makers, bureaucrats and politicians who insist on shore front "development" under the banner of demands, and environmental scientists and NGO's who insist on basic ecological and environmental principles. Therefore, there is a vertical split with perceptible differences of opinion when coastal zone issues are to be tackled.

The builders lobby, real estate developers and resort owners are the leading players in the present construction activity. Consequently, the weaker sections of the society, for example, are being pushed away from their traditional lands and even professions as traditional fishermen loose space for their age-old fishing activities. Considering charges of corruption against government department officials and innumerable infringements, which are going unchecked in the coastal zone, the builder - politician nexus is suspected.

With various such problems in view, and since uncontrolled development had started along the coastal zone of the country, the Ministry of Environment and Forests (MoEF), New Delhi, enacted a legislation called the Coastal Zone Regulation (CRZ) Notification (Anonymous, 1991) issued under the Environment Protection Act of 1986. The main purpose of this notification, which lists various rules and guidelines, was to control and minimize environmental damage to coastal stretches including estuaries and backwaters. No Development Zones (NDZ) which had to be measured from the High Tide Line (HTL) were notified for beaches, rivers and backwaters. Accordingly, all coastal states of India were bound to formulate coastal zone management (CZM) plans classifying coastal stretches as coastal regulation zones, based on the nature of existing ecosystems and/or development. The CZM plans (Anonymous, 1996a) prepared by the state of Goa were rejected (Anonymous, 1996b) as the document contained some flaws. In the meantime, flagrant misuse of coastal spaces on a national scale forced some environment conscious citizens to file a public interest litigation based on which the Supreme Court of India validated the CRZ Notification on 25 April 1996. In June 1996, The MoEF requested the National Institute of Oceanography, Goa, for information on the eco-sensitive areas of coastal Goa. After a detailed survey, all related data were submitted to the MoEF, in August 1996, in the form of an unpublished report (Anonymous, 1996d). The final approval of the plans, with certain modifications to be incorporated, was issued in September 1996 (Anonymous, 1996e). This is the prevailing law, which governs developmental activities along the coasts, rivers and backwaters of Goa.

Despite clear legislations in place, it is observed that a large number of structures presently found in sandy areas close to the beach tantamount to blatant violations of CRZ regulations. Many constructions have been stopped by the courts (Anonymous, 1997c; Anonymous, 1998a), some judicial cases are pending, some structures were demolished but were subsequently rebuilt in the same place. Different types of infringements in the coastal zone and their legal implications have been documented elsewhere (Alvares, 1993). Recently, based on site visits (state government committee) in 1996, more than 20 luxury hotels and resorts were issued show cause notices for violating CRZ regulations, a fact admitted by the state legislative assembly.

Therefore, following complaints from citizens and NGO's, the first writ petition was filed in 1988 and have continued since then. Courts have intervened in coastal matters for a variety of reasons as, in some cases, constructions started before obtaining requisite clearances. Letters from concerned individuals have been converted into writ petitions. Similarly, courts have ordered hotels to remove all structures within 200 m area, including houses, rooms, wells, roads, fences. The state government was also ordered not to allow constructions within 200 m from HTL (Anonymous, 1996c). Courts have also ruled that disturbing sand dunes threatens coastal ecology and hence sand mining leases had to cancelled. Stay orders have been imposed in several cases, the latest being a resort in Baga and a construction in Utorda (Anonymous, 1997,c; Anonymous, 1998,a).

The CRZ legislation has apparently hurt the powerful builders lobby, resort owners and influential members of the government some of whom had already acquired large tracts of prime coastal lands (Anonymous, 1997b; Anonymous, 1997d). Thus, the CRZ Notification was sought to be criticised and misinterpreted to suit vested interests, as such defeating the very purpose for which it was enacted. The outcome is what we see today (Figure 12.1): an unprecedented and haphazard growth of coastal structures with consequent alterations of the face of the coast.

HUMAN INTERFERENCES

Reference to ancient documents, maps and planning literature, supplemented by discussions with concerned individuals offer valuable information about the past. The effects of development on the coast can be assessed by comparing the coastal configuration and topography prior to major human occupation with conditions that exist today; this approach has been attempted wherever possible. In Goa, to our knowledge, only isolated data on human intervention on the coast is available, and as such, we do not have comprehensive details about the present scenario along the entire coast. Major factors responsible for coastal degradation are described in the sections to follow.

1. Impact of construction activities on sand dune fields

Since the 1970's, heavy construction activities and associated infrastructure requirements of the tourism industry have led to uncontrolled and haphazard growth of many coastal stretches as several coastal belts are overbuilt. The northern Baga-Candolim strip is a classic example as, at many points there is no proper public access to the beach. Here, the entire coast has turned into a continuous row of hotels and resorts (Figure 12.1). It is therefore obvious that these coastal areas have crossed a saturation point (Mascarenhas et al., 1997; Sawkar et al., 1998; Rosario, 1998). Whereas Colva and the Cavelossim-Mobor areas of central Goa will meet the same fate in a short time, Palolem in the south will follow very soon.

Compared to all coastal geomorphic features, sand dunes have borne the maximum brunt and have suffered the greatest from indiscriminate anthropogenic pressures (Lobo, 1988; Mascarenhas, 1990, 1996a & b). Large scale conversion of dune belts has taken place, and thus, the functions of a dune ecosystem virtually changed to a residential area. These effects are identified wherever man-made structures have encroached on dune fields (Figures 12.2 and 12.3). These impacts can be observed along most of the sandy coast of Goa, particularly from Baga to Mobor including the two major estuaries in between.

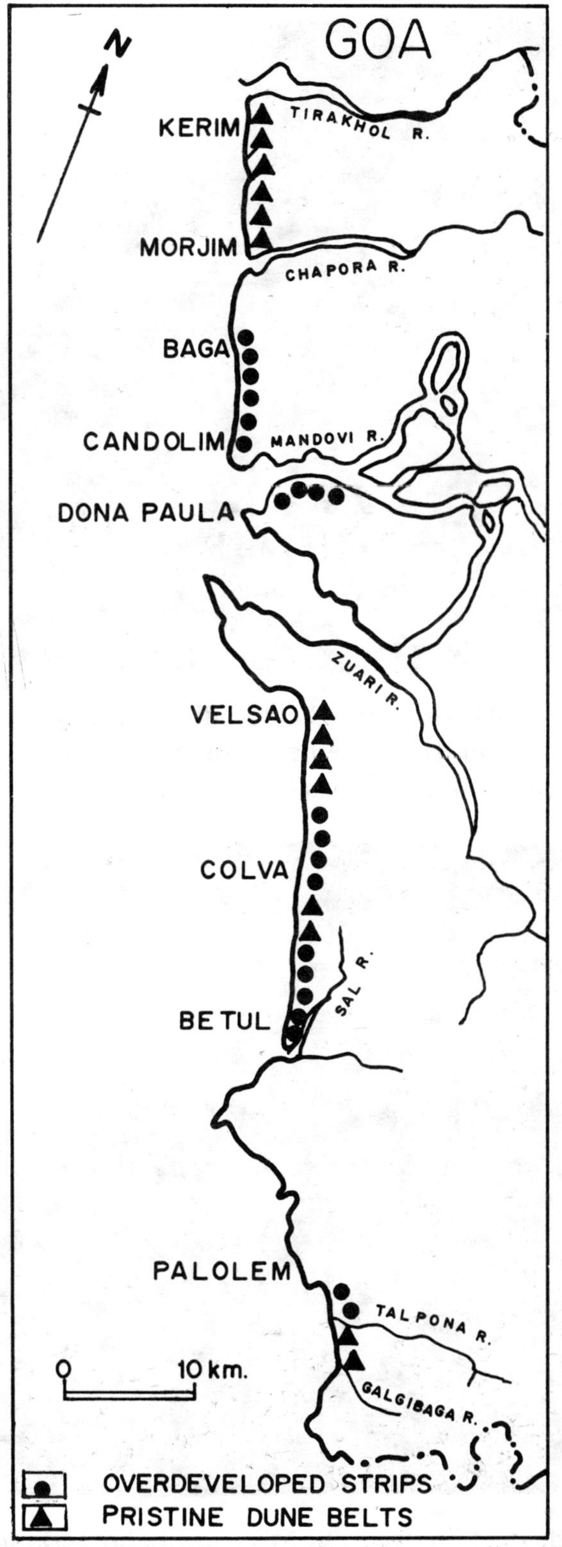

Figure 12.1: Map showing coastal strips which are overbuilt and areas where pristine dune belts can still be found.

Figure 12.2: A resort which was being built on the dunes at Baga, less than 50 m from HTL. Being a violation of CRZ, the project was stopped by the courts (Photo: August, 1996).

Figure 12.3: Residential dwellings at Miramar being built on a dune complex that was completely levelled; the former dune system is now a residential area (Photo: May, 1992).

Our studies along all dune belts of coastal Goa clearly show that high dunes have been razed, levelled, flattened or simply removed so as to make place for hotels, resorts or other structures. Large dune areas are thus found obliterated and damaged due to human intervention. All these anthropogenic activities have resulted in irreversible alterations to the dune systems as removal of large volumes of sand has created a negative sand budget, which in turn has affected their stability. Many parts of the coastal zone have changed from virtual wilderness to areas where buildings arc mushrooming, and as such, many of the coastal stretches bear little resemblance to the primitive shore that existed before.

Several structures have been identified on the beach - dunes along the HTL, the most prominent being a resort at Miramar and many other smaller ones at Anjuna and Palolem. Excavations and removal of dune sand close to the beach have created deficiencies in the volume of sand. All these sites are precisely the ones where significant erosion has been noted. Although the erosive processes are localized at the moment (except at Miramar), it is found that these hard structures which are in direct contact with marine water are mainly responsible for shoreline erosion, as clearly observed at Palolem. Pertinent details are discussed in later sections.

2. Impact of dune sand extraction

Dune sand deposits have been under constant pressure for various reasons such as:

(a) the common man uses sand (on a low scale) as cement mixture for houses and homes,

(b) industrialists and builders plunder sand which is used as a raw material for a variety of construction purposes,

(c) all coastal resorts and hotels located in sandy areas find sand as a readily available resource, and

(d) levelled dunes can be converted into plots for further development.

Permission was required to mine sand during the erstwhile Portuguese regime. After liberation in 1961, sand mining leases were granted to private parties for as long as 20 years (Alvares, 1993). Even during the late 1980's, there were sand mining sites at Anjuna, Baga, Candolim, Arossim, Benaulim, Cavelossim and Mobor. As a result, thousands of tones of sand were extracted and even exported outside the state (Alvares,

1993). Following protests from people and expert advice from the scientific community, in 1983, the state government issued orders cancelling all leases. Nevertheless, sand extraction continued unhindered till 1988 when the courts intervened decisively forbidding sand mining. Our observations however show that clandestine sand extraction still continues (Figure 12.4).

The dune sand extraction has shown many impacts in the form of danger signals (Lobo, 1988; Anonymous, 1997e). Saline water entered paddy fields in the hinterland at Mobor; fruit bearing trees were destroyed. In 1977, a dune which was removed left a scarp/gap through which storm winds blew tonnes of sand onto the hinterland paddy fields affecting paddy output and fruit trees which were scorched by salt air (Alvares, 1993). At Varca, large amount of sand from dunes over a wide area were plundered by contractors to supply sand for constructions that made the area flat and featureless. In 1977 and 1978, disturbances of dune systems resulted in a storm surge in Mobor, and attack by storm waves and inundation in Majorda, Betalbatim, Varca, Carmona, Cavelossim (Lobo, 1988). The elimination of dunes has lead to levelled areas, which are later converted into plots. Thus, dune sand extraction has shown disturbances in the form of erosion and salt water invasion inland as such anthropogenic activities work against the sound ecological principles of the coastal zone, the dune systems in particular.

3. Effects of roads on sand dunes

Access roads are seen at many places along the coast, being more concentrated on the dunes of Baga - Sinquerim sector, the rest being found in the south. Roads over the dune belts have lead to obliteration of dune topography (Mascarenhas, 1996,a), removal of sand, levelling of dunes, desecration of dune vegetation which acts as sand binders, all of which generate loose and free sand resulting in its excessive mobility. In all cases therefore, such roads are covered, sometimes in totality, by a layer of sand; this is a recurring process. Such a phenomenon is very evident wherever roads occupy former dune fields, and very conspicuous at Calangute and particularly the parking lot in Colva. In fact, each and every approach road to the beaches is covered with a layer of sand.

The Miramar circle, which was built about more than 100 years ago, can be cited as a human failure. During windy days in summer, large quantities of dry sand are blown inland. The beach, which is excessively frequented, is devoid of dune vegetation, and therefore, bare in most of the areas. The result is that the Miramar round-about and the radiating roads are covered by sand for most part of the year, creating dangerous road conditions for traffic. Imbalance in the sand budget has rendered sand mobile, and comes to occupy its rightful place: the circle, now located on former sand dunes.

The new coastal road to Dona Paula will meet the same fate. A major part of the road is laid on dune sands. It is already evident, at least at two places, that sand is attacking the road where small accumulations of sand are noticed, although on a low scale. If remedial measures are not taken, it is obvious that this problem will aggravate further and the road (as in other areas discussed above) may be drowned in sand, thus creating hazards to drivers in general and motorists in particular.

4. Impact of beach shacks

Beach shacks comprise simple huts, made of bamboo and coconut tree leaves, are of a purely temporary nature. In the past, there were few of them but at the moment, the ever increasing competition has resulted in their proliferation as almost 220 shacks (and numerous unauthorised ones) have been granted licences for about 60 km of beach length (D'Mello, 1997). Strictly speaking, since a sandy beach falls under CRZ-I category, even a temporary stall such a shack cannot be allowed on the beach according to the CRZ notification. However, since shacks are seasonal and form the livelihood of hundreds of unemployed individuals, it appears that the beach shacks have come to stay.

The profile of a sandy beach keeps changing and adjusting itself towards an equilibrium, as a beach generally restores itself. Therefore, beach shacks, which come and go are not expected to induce any large-scale damaging effects from the environmental point of view. However, their rising numbers have resulted in severe crowding which is gradually leading to over exploitation of limited beach space. As they are frequented by a large number of people, shacks generate huge quantities of garbage, which unintentionally mixes with sand, and hence litter large areas, and invariably finds its way into the estuaries.

Shacks are bound to occupy only the upper part of beaches; but many of them are seen beyond these limits. The adjacent sand dunes are thus invaded, as seen in Candolim-Calangute belt. This is where ecological degradation takes place as dune vegetation is uprooted and cleared to make place for them, an environmental damage that can be significant and sometimes irreversible. This is also because sand dunes are flattened and as such loose their intrinsic value. Therefore, beach shacks, although simple in design have not only flattened dunes and added litter onto the beach, but have their own share towards coastal degradation.

5. Influence of recreation on dunes

Activities of pleasure seekers and picnickers are not restricted to a particular area, but can be observed almost along the entire coast particularly where sandy stretches as the dunes are found. Two dune areas, in Baga and in Candolim, are regularly used as a football ground by the villagers and also tourists. Although these recreational activities may not be termed as harmful to the dune environment, it does create ecological degradation in areas which are most sought and frequented. Driving on the beach which is an illegal activity is often noticed, parking of vehicles on dunes (Figure 12.5), play fields on dunes, continuous movement by pedestrians and cyclists are factors which destroy dune vegetation, flatten dunes, induce shifting of sand and renders the sand mobile, thus affecting the stability of sand dunes.

6. State of dune vegetation

The status, importance, uses, types, ecological aspects and management implications of dune vegetation have been studied earlier (Desai, 1995). Dune vegetation, an integral part of the dune ecosystem, plays a crucial role as the vegetal cover acts as an effective sand binder and hence promotes the stability of dunes by reducing erosive processes. Dune vegetation is vulnerable to human interference due to its fragility. Our observations have revealed that sand dune vegetation is in various stages of degradation in different parts of coastal stretches. The maximum damage appears to be along the Baga-Candolim strip.

There are innumerable anthropogenic factors which are responsible for the destruction of dune vegetation: real estate owners who prepare plots for buildings, hotels or resorts clear all dune vegetation; roads are laid over levelled dunes thus eliminating the vegetal cover; beach shacks and various other recreational activities have shown similar effects as dunes are rendered bare. Nevertheless, there are several isolated tracts along the coast where sand dune vegetation is still found in its pristine form as in the northernmost coastal strip of Goa.

7. Litter on beaches

Litter found in sandy areas is attributed to various sources. With the ever increasing pressure on the coastal belt, and as more people are moving towards the coast, our beaches are getting dirtier. The Baga-Candolim stretch, Miramar, Colva and even Benaulim can easily be singled out. For instance, the annual sea food festival held at Miramar produces heaps of garbage, a part of which gets mixed with the sand. It was

Figure 12.4: A huge sand dune near Miramar has been mined and sand used for constructions in the vicinity; a building now stands in the place of the dune (Photo: May, 1992).

Figure 12.5: A large area on the dunes being used for parking all types of vehicles at Candolim. The dunes are now flat, bare and devoid of vegetation (Photo: January, 1998)

Figure 12.6: Brutal cutting of a hill slope for road widening for the bridge approach roads at Betim; over the years, the loose mud has deposited in the Mandovi river creating shoals at places (Photo: December, 1997).

Figure 12.7: A site of vigorous erosion where the original beach at Campal has vanished. A resort is being built at a place prone to erosion (Photo: January, 1998).

noticed that a large pit in the dunes at Calangute is used to dispose garbage by the public at large. At Colva, the prominent creek is used as a dumping site for all types of waste by stall owners in the vicinity. Hotels in south Goa produce large dumps of non-biodegradable garbage which along with sewage is discharged in the river Sal. Interestingly, large rafts of plastic waste were found floating in the river Sal; a close look revealed plastic items (and even seals) dumped by starred hotels and not by the locals.

Litter scattered along the coast ultimately finds its way into the rivers or the sea and is subsequently thrown back on the beach as can be observed along most beaches during the post monsoon period. According to newspaper reports, the Navy identified large rafts of plastics floating in the sea off Goa. Therefore, irresponsible disposal of garbage in the coastal zone has resulted in a progressive build up of plastic waste in our estuaries.

8. Salt water ingress

Till a decade ago, there were no complaints of salt-water intrusion into coastal aquifers. Problems started in recent years when a large number of bore wells with powerful pumps were installed within the 200 m limit, notably by the builders, the hotel industry and also by private parties (Alvares, 1993). It is to be noted that the pumping of underground fresh water in the dune belts is forbidden by CRZ regulations particularly within 200 m from HTL. Beyond this limit, permission of competent authorities is needed. Despite this, several infringements are reported, as large number of cases were detected beyond 200 m where borewells are regularly used to draw water particularly by the hotel industry (Alvares, 1993). In some coastal places, water is openly sold by private parties to hotels for use in their swimming pools. Thus, this ban is routinely violated. Since domestic water supply is grossly inadequate to meet the rising demands of hotels for their lavish use, no action is taken on builders and hotel owners for tapping underground water for consumption.

This kind of unchecked exploitation of ground water has resulted in lowering the groundwater table and seepage of saline water into fresh water wells. In 1992, several wells of Mobor peninsula were invaded by salt water (for details see Alvares, 1993). Coastal acquifers in north Goa have been affected by ground water fluctuations and seasonal increases in salinity, as wells near the sea shore show influence of salt water encroachment as it is hydraulically connected to coastal water table; therefore ground water quality is deteriorating (Chachadi and Kalavampara, 1995).

The outcome of deliberate tampering with coastal acquifers under dune systems is distinctly demonstrated in south Goa. A resort located between the sea and the river excavated a huge dune area to be converted into an artificial sweet water lagoon without a proper assessment and despite repeated warnings against this adventure. To their dismay, saline water trickled, attacked and filled the artificial lake and today we only find a salt water lagoon. Such an unwanted human interference with the dune system is a clear example of salt water ingress into a former fresh water acquifer.

9. Human activity on coastal hill tops and slopes

The coastal strip of Goa also consists of evergreen hills and promontories with wooded or bare hill slopes which often intersect beaches and jut into the sea. Forests along hill slopes offer an intrinsic natural beauty and also contribute towards slope stability. Coastal hill slopes which control erosion and also recharge acquifers are integral parts of the coastal zone ecosystems.

Of late, we find lot of building activity on plateaux and coastal hill slopes. Frenzied development can be observed along the hill slopes of Dona Paula-Bambolim stretch, along the river Zuari as well as within the Mandovi estuary, on the banks and slopes opposite the city of Panjim. A large number of constructions have gradually taken over and encroached the coastal hill slopes, some of which have almost reached the water line, whereas others, perched on the edge of the plateau have constructed steps of concrete which lead to the water. In addition, hill slopes at many other places within estuaries have been brutally sliced for widening of roads as in Betim (Figure 12.6) and laying of railway tracks as in Cortalim, Talpona and Galgibaga.

The environmental impacts of such actions can be manifolded. Human activity along hill slopes has lead to mud flows into the rivers resulting in heavy siltation at places. This is clearly evident on either side of the railway track at Talpona estuary. Construction activities are not only spoiling coastal aesthetics but also gradually leading to the degradation, deforestation and concretisation of coastal hill slopes. Shoals and mud flats which are promptly colonised by mangroves within rivers are a direct consequence of negative impacts due to interference of man along these slopes. Coastal hill slopes, particularly the wooded ones which recharge acquifers and arrest erosion, need to be protected from any development whatsoever.

10. CRZ violations

Infringements in the CRZ is a burning problem in Goa, despite clear rules and guidelines. The notification has been deliberately misinterpreted to suit individual interests. The result is seen in the form of blatant large scale violations of CRZ regulations observed along the open sea front as well as along the rivers and backwaters. The sanctity of this unique regulation is therefore at stake due to the folowing activities:

Buildings and constructions: Based on our field observations, it is clear that there are numerous irregularities along coastal sandy stretches. The setback lines as stipulated in the notification have been conveniently set aside (Anonymous, 1997c; Anonymous, 1997d). There are many such cases in the Baga-Candolim sector; the Calangute panchayat has been hauled up for the maximum number of environmental violations (Anonymous, 1996c). Miramar-Campal area also deserves mention (Figure 12.7). During the last few years, south Goa dune belts are under assault. The Mobor peninsula which was once declared as a green cover area, has witnessed a frenzied proliferation of resorts which have violated almost all rules stipulated in the environmental guidelines (Alvares, 1993; Anonymous, 1997b). Our observations show that the latest negative developments are in Palolem where a number of small buildings have mushroomed close to the beach, even on the HTL. A concrete ramp to facilitate vehicles on the beach is probably an unprecedented infringement along the beaches of Goa.

Sand extraction: Although prohibited by law, unscientific clandestine dune sand extraction which has gone unchecked is rampant at several sites along the sandy coasts (Anonymous, 1997e). A large dune near Miramar was almost completely mined as recently as 1992 (Figure 12.4).

Bore wells: Since piped water supply is often irregular and does not meet the existing demand, many real estate owners, the hotel promoters in particular, resort to unfair practice of ground water tapping by sinking bore wells (Alvares, 1993). Although several defaulters have been booked, many such cases go unnoticed and clandestine operations such as selling ground water continue unabated.

Roads: In some cases, ancient paths used by beach users are slowly being converted into roads some of which are already tarred, whereas others are incomplete and/or stopped by legal means.

Fences/public access: Several promoters of hotels and resorts have put up fences in the no development zone, sometimes up to the HTL. Existing law does not allow fences or walls within the 200 m zone and yet they are frequently seen, at Calangute for example. Courts had to intervene in some cases so as to restore public access to the beaches. Several resorts in north and south Goa and a luxury hotel in Dona Paula can be cited as examples.

DISCUSSION

The relationship between human activities and coastal changes and evolution is not fully understood (Nordstrom, 1994a and 1994b). There appears to be a lack of scientific interest in the evolution of developed coasts because geomorphologists do not study human altered coastal systems, but mostly deal with landscape evolution at larger time scales. Nevertheless, there are several reports on the effects of man made structures on the coastal zone ecosystems world wide. We can also find field oriented studies on the effects of buildings, dwellings and roads on the adjacent coastal beach-dune systems as well as impacts of tourism and allied activities on the coastal morphology. The present study mostly aimed at coastal sand dune ecosystems is therefore one such initiative in this direction.

The anthropogenic impacts discussed above clearly indicate that construction of resorts and buildings, dune sand mining and roads on sandy strips are the three major factors responsible for the large scale degradation of coastal sandy strips. In addition, flagrant violations of prevailing legislations have complicated matters further as changing coastal configuration has remained unchecked. The consequent result is the elimination of coastal sand dunes; this is the single largest irreversible impact that is being faced at present.

Impacts such as the marine salt water ingress which is seasonally showing up in coastal acquifers is attributed to excessive ground water withdrawal; although somewhat under control, this effect could lead to alarming problems if these trends continue. In comparison, all the remaining man-made effects related to beach-dune systems can be termed to be minimal as, in most cases, these impacts can be fully reversed.

The main prerequisites for the formation of sand dunes are wind, sand and vegetation - three elements with complex and dynamic interactions (Carter, 1988). Wind plays the most important role; its direction, frequency, duration and speed are to be considered as only winds with a speed greater than 16 km/hour can lift, displace and transport fine dry sand (Paskoff, 1994). This is more so along low gradient beaches particularly when they are sub-aerially exposed at low tides, and when strong winds blow perpendicular to a particular coast. Vegetation behind the beaches also play a key role as these objects act as wind breakers which force wind to drop sand along its path; vegetation thus traps and stabilises moving sand (Carter, 1988; Paskoff, 1994).

The first step in the formation of a dune is a berm which is an accumulation of sand brought up by the waves on the beach at a point just above the highest high tide. The berm swells due to the eolian import of sand. Thus, these small sand mounds which form, finally develop into a continuous chain of sand dunes. These eolian bedforms can be symmetrical, several meters in height and can differ from place to place. Their dimension depend on the degree of evolution and the aptitude of vegetation to trap and retain sand. The destruction of sand dunes in Goa is almost exclusively anthropogenic (Figures 12.2, 12.3, 12.4 and 12.5). This can be clearly observed in Baga-Candolim where resorts and hotels are dense, Miramar-Caranzalem, and Betalbatim-Mobor, but more particularly at Cavelossim. Although they had remained free from human intervention, they are now too frequented subsequent to the promotion of coastal tourism. They have thus suffered greatly from the assaults of tourism and other related activities which have spread over coastal spaces during the last two decades or so. Their existence is therefore in peril: destabilisation due to the degradation of their vegetation which is sensitive to the human presence, flattening due to sand extraction, or their destruction and elimination due to the creation of plots by the sea side. Their disappearance is total where sand is being extracted for construction activities, and also where dunes have been simply levelled and flattened to erect buildings and resorts.

Similarly, the destruction of dune vegetation (Figure 12.5) is crucial as sand binders appear to be exclusively anthropogenic and attributed to several factors discussed above. In several areas, the frequent visits of picnickers, tourists, revellers, vehicles, cycles have rapidly degraded sand dune vegetation which is particularly sensible to human interference. Native plant species are prevented from colonising because of the preference for the lawn grass and exotic shrubs for landscaping. Real estate developers and owners of resorts have replaced natural vegetation with gravel, cemented paths, or simply maintained an unvegetated barren surface which is used for other purposes such as parking (Figure 12.5). Natural dune vegetation can only be seen on existing stretches which have so far remained free from development.

Littoral sand dunes which constitute spaces of recreation complementary to the beaches behind which they are located, are edifices of extreme fragility and are essential for the sedimentary and ecological equilibrium of coastal environments (Carter, 1988). Their destruction can sometimes be attributed to natural causes. If the vegetation is destroyed by prolonged dry periods, or other reasons, they are destabilised as they are laid bare. If subsequently, a dune is cut by high waves leaving an escarpment, the latter is capable of increasing wind speed manifold. Wind then opens cavities through which free sand is liberated and bombarded inland thus attacking populations and buildings located along its path.

Coastal dunes show a dynamic natural behaviour which has to be to be understood, respected and not contradicted, as is often being done, by anthropogenic influences (Paskoff, 1994). Dunes cannot simply follow a pattern which is not naturally theirs. Thus their intrinsic value is lost (Nordstrom, 1990), and as a result, consequences are noticed on the beaches with which they are associated (Carter, 1988). A sand dune belt has multiple functions and hence of immense value to coastal populations as mentioned below.

i. The beach-dune environment is a highly organized system, features of coastal stability, the result of a delicately balanced ecological equilibrium between the forces of the ocean and loose coastal sediments,

ii. The dune environment is classified as edifices of extreme fragility, sensitive and vulnerable due to its propensity for changes under even slight environmental stress,

iii. Sand dune chains, as high as 10 m in Goa, are categorized as Nature's line of defence as they arrest blowing sand, deflect wind upwards, assist in the retention of fresh water and protect the hinterland from attack by waves,

iv. Coastal sand dunes serve as "sand banks" (Paskoff, 1994), are sources of beach nourishment and also neutralize and dissipate wave and current energy in the coastal zone and maintain the sedimentary and dynamic equilibrium of the dune-beach ecosystem,

v. Dune vegetation acts as sand binders and preclude loose sand from advancing inland and thus menacing coastal populations,

vi. Dunes protect the hinterland from winds and other forces and hence make the zones behind dunes as areas of peace and tranquillity,

vii. Sand dunes of Goa are about 5000 years old (Sriram and Prasad, 1988) but can be as old as 6400 years (Kale and Rajaguru, 1983) and

viii. The predicted sea level rise is bound to inundate coastal areas with a profound effect on the coastal zone; dunes stand guard against any sea level rise as they will act as Nature's wall of defence. Therefore, these geomorphic edifices have to be preserved at all costs.

It must be mentioned that sand dune ecosystems were not included within the ambit of CRZ-I in the original notification. Since sand dunes are sensitive and fragile geomorphic features that are vulnerable to even small environmental stress, and since they act as Nature's line of defense against the forces of the ocean, a case was made for sand dunes fields as CRZ-I (Anonymous, 1996d; Mascarenhas, 1996b). This was following the request made by the Ministry of Environment and Forests (in June 1996), to the National Institute of Oceanography, Goa, to identify areas which are ecologically sensitive. Accordingly, a report was submitted to them in August 1996 (Anonymous, 1996d). Based on this report, and considering the ecological importance and immense use of sand dunes to society, the Ministry of Environment and Forests declared and classified the sand dunes belts under the CRZ-I category. Subsequently, in September 1996, the Ministry issued separate orders (Anonymous, 1996e) categorically directing that sand dunes are to be independently included under CRZ-I, a parameter that had to be incorporated in the coastal zone management plans of Goa. Therefore, in coastal areas characterized by sand dunes, no new constructions are permitted up to 500 m from HTL.

The damage and destruction of dunes, irrespective of the agents responsible, pose various consequences for the coastal environment (Miossec, 1988; Nordstrom and Gares, 1990). Disrobing a sand dune results in the formation of free sand which is rendered mobile; sand therefore keeps shifting (Figures 12.4 and 12.5). Along the coast of Goa, the transfer of sand landward far into the hinterland is a common phenomenon observed particularly during windy days at several places; the Baga-Candolim sector is one such example. The risks of inundation are increased when sand is removed on a large scale. Excessive pumping of fresh water leads to salt water ingress and contamination of coastal acquifers as identified at Candolim (Chachadi and Kalavampara, 1995). Elimination of dunes also induce erosion, as beaches are starved of the requisite sand budget (discussed below). All these factors adversely affect coastal equilibrium and hence its stability.

Research has revealed that recreation can have adverse effects on dunes (Vogt, 1979). Coastal buildings alter wind flow around them, reduce the velocity on the lee; structures built on dunes induce localised deposition and cause local increase in the height of the dunes and reduce the inland migration of the dunes. Even a moderate development on dune belts strongly affect sediment transfers (Nordstrom and McCluskey, 1985). Buildings also change the location of accretion and scour on beaches and dunes. Large buildings cause a reversal of regional wind flow; wind blown sand accumulates against buildings, houses, boardwalks and roads creating a diverse dune scape, distinctly different from the one created under natural conditions (Nordstrom and Jackson, 1993; Nordstrom, 1994b). Sand accumulation on the roads can be seen at Miramar. In our case, the major drawback is that the human intervention on the sand dunes of Goa has never been quantified, an exercise which requires immediate attention.

Since sand dunes are eolian bedforms which are dynamic and mobile, they should be allowed and be able to evolve freely and naturally in form and space. Research has shown that dunes in certain areas have displaced and migrated naturally by tens of meters towards the sea as well as away from it as compared to their primitive location. This has been ascertained by successive aerial pictures (Psuty, 1988; Paskoff, 1994). Since a natural sand dune is in a natural equilibrium with the beach, if the beach retreats the dune should be able to do likewise (retreat) so as to assure its role as a sand reservoir in order to maintain the equilibrium between the dune-beach ecosystem. Since a majority of beaches are in a process of retreating due to a slow rise in sea levels, the dunes should also retreat landward by rolling on themselves. Such a mobility and flexibility of sand dune are rarely considered. For this, ample space is necessary and is to be reserved for migration and sand dunes should be at liberty to do so. The present style of development as observed along the sandy stretches of Goa, therefore, works against the natural behaviour of sand dunes. That was precisely the reason why our ancestors never occupied sand dunes, built houses and roads away from beaches and always lived behind sand dunes, a system that can still be observed along the entire west coast of India. Their wisdom to maintain a perfect coastal equilibrium cannot be questioned.

Erosion of beaches and dunes has been observed at several places along the open sea front and within estuaries. This phenomenon, although localised, is significant at Arambol, Anjuna, Majorda and Colva. A proper scientific proof pending, these events are attributed to annual changes that a beach undergoes as these areas regenerate themselves during the post monsoon period, as has been observed by the author over the last few years. Sea erosion is however most pronounced at Velsao and Agonda as evidenced by a large number of tall uprooted trees (Mascarenhas, 1996b). These are areas of periodic intense wave energy and heavy erosive tendency. These are instances of natural erosion.

Near Verem in the Mandovi estuary, the existing small sandy beaches are gradually disappearing. This is attributed to a series of walls erected recently, mostly by hotels. Similarly at Palolem, many hard structures have been built along the beach. This human intervention has resulted in shoreline changes, wherein the sea is advancing landward as observed over the last two years. A remarkable erosive activity is proved by breaches in concrete embankments and uprooted trees; marks left on these walls show a lowering of the level of the beach indicating that a large volume of sand has been washed away. The erosive activity at Palolem, attributed to human action, is definitely and undoubtedly irreversible.

The most alarming erosive processes are seen along the Miramar-Campal sector (Figure 12.7). The entire stretch which constituted a dune belt with a thick vegetal cover as late as the 1950's has now ceased to perform the functions of a beach dune-ecosystem. The dune complex no longer exists; buildings have come up in its place. A stone embankment was erected during the last decade; but erosion continues with full vigour as hardly any sand can be seen along this structure. The lighthouse which was once on the shore is now found in water and is hence under threat as it faces an imminent collapse. This is an example where the planners got the taste of their own medicine as the stone wall keeps collapsing. This is because hard structures increase the turbulence of waves resulting in the removal of sand (Kraus, 1988; Frihy et al., 1996); the beach at Campal has therefore disappeared. Despite the fact that the shoreline is prone to erosion, a huge hotel complex is now under construction on a dune at Miramar, precisely at a place where it should not have been (Figure 12.7). This is a classic example of a human blunder.

The coastal hill slopes and even promontories are unfortunately not covered by the CRZ Notification, although plateaus are ideal for future development. However, wherever buildings are coming up, leaving a reasonable setback from the edge of the cliff or hill slope is imperative, be it from the safety or environmental points of view. But in reality, this rule is not observed. The hill slopes are being invaded with impunity. If this trend is not reversed, constructions along coastal hill slopes are bound to assume alarming proportions in future.

Along western India including Goa, the rise in the eustatic sea level over the past decades is estimated to be 1 to 1.5 mm/year, based on tide gauge records along the west coast (Subrahmanya, 1996). Although a nominal rise in sea level will not affect population and structures at higher altitudes, it will have a profound influence on flat beaches, estuaries and lowlands. Such an effect may be less severe on the sandy beaches of Goa as most of them have an appreciable gradient and, as such, the dunes are situated at a significantly higher level, beyond the reach of normal sea waves. However, in the event of a (predicted) one meter rise in global sea levels during the next century, the erosive capacity and storminess of the seas will increase manifold (Haq and Milliman, 1996). Such a scenario will eventually invade and erode beaches, the flat ones in particular, as large volumes of sand are expected to be washed away by stormy waters. Beach erosion is induced by rising seas which shifts wave action to progressively higher levels, leading to extensive and rapid erosion. This will in turn influence and modify the sand dune systems (Carter, 1988) as a large part of beaches may disappear. Therefore, these precautions ought to have been included in the coastal zone management plans as rows of beach resorts are located along the coastal belt. This important criteria is however conveniently omitted.

This is what brings us to the question of setback lines. Considering the fragility and mobility of sand dunes, it is imperative that NDZs be strictly enforced. Buildings should be located behind sand dunes. Free spaces are also needed along the edges of coastal plateaux which are ideal for future development. Similarly, in the event of a predicted sea level rise, the most cost-effective long term solution is to set aside land for future marine transgressions. Hence the imperative need for reasonable set back lines.

CONCLUSIONS

- Laxity in implementing and enforcing legislations has resulted in unauthorised structures and uncontrolled growth along the coastal zone. Blatant large scale violations of CRZ regulations are observed along the sandy stretches of Goa. Some structures are seen even along the HTL.
- Our study reveals that (a) construction of resorts and buildings on dunes, (b) mining of dune sands and (c) roads on sandy strips, are the three major factors which are responsible for the coastal degradation.
- Sand dunes have borne the maximum brunt and have suffered the greatest from indiscriminate anthropogenic pressures which mostly include haphazard construction activities. Pristine dune belts do however exist.
- Localised coastal erosion is observed wherever hard structures interfere with sea water. Such man-made problems are generally irreversible.

Mandatory setback lines are imperative. Sand dunes are landforms essential for coastal stability, they are ecologically sensitive, they are sometimes mobile and hence shifting in nature; the threat of an impending sea level rise will make dunes act as Nature's line of defense. These factors collectively demand that coastal dune systems be compulsorily designated as no development zones.

ACKNOWLEDGEMENTS

The author is grateful to the Director, NIO, Goa, for permission to publish this paper. He also expresses his gratitude to Dr. V.K. Sharma, Indira Gandhi Institute of Development and Research, Mumbai, for the invitation to contribute this article. Dr. K. Sawkar and Mr U. Sirsat, NIO, are thanked for their help in the field.

REFERENCES

Alvares, C., 1993. Fish, Curry and Rice: A Citizen's Report on the State of the Goan Environment, Ecoforum, Goa, pp. 260.

Anonymous, 1987. Master Plan for Tourism Development in Goa, Department of Tourism (Draft Report), Government of Goa, July 1987, 130 pp.

Anonymous, 1991. The Gazette of India, Notification, S.O. no. 114(E); The Ministry of Environment and Forests, February 20, 1991.

Anonymous, 1995. State files affidavit denying CRZ violations. Herald, 18 August, p. 1.

Anonymous, 1996,a. Coastal Zone Management Plans for Goa, Town and Country Planning Department, Government of Goa, pp. 69.

Anonymous, 1996,b. Delhi rejects Goas's coastal plans for constructions. Herald Illustrated Review, 96:1.

Anonymous, 1996,c. Court passes strictures against Calangute panchayat: illegal constructions within HTL. Herald, 10 April, p.1.

Anonymous, 1996,d. Comments on the Coastal Zone Management Plans of Goa. Report submitted to Ministry of Environment and Forests, New Delhi. National Institute of Oceanography, Goa , September, 1996.

Anonymous, 1996,e. Coastal zone management plans of Goa (approval notification). Ministry of Environment and Forests, No. J-17011/12/92-IA-III, September 26.

Anonymous, 1997,a. Sea turtles on the run as humans take over beaches. The Navhind Times, 20 February, p. 1.

Anonymous, 1997,b. Gross violations of coastal zone notification at Benaulim. Herald, 30 March, p. 8.

Anonymous, 1997,c. Supreme court upholds high court order on Piva resorts. Herald, 3 May.

Anonymous, 1997,d. Government allowing resort to come up on the bank of River Mandovi? Herald, 8 September.

Anonymous, 1997,e. Sand extraction threatens south Goa beaches. Herald, 21 November.

Anonymous, 1998,a. Magistrate halts building work near high tide line. Herald, 1 February.

Anonymous, 1998,b. The master plan for tourism: fact or fiction? Herald, 5 March.

Carter, R.W.G., 1988. Coastal environments: an introduction to physical, ecological and cultural systems of coastlines. Academic Press, London, 607 pp.

Chachadi, A.G. and Kalavampara, G., 1995. Degradation of groundwater environment due to sea water intrusion in parts of Goa coast (Abstract 9.3). XII Convention, Indian Association of Sedimentologists, Goa University, p. 83.

Chari, B., 1998. Blind tourism policy encouraging illegal constructions, Herald, March 5.

Desai, K., 1995. The structure and functions of the sand dune vegetation along the Goa coast. Thesis, NIO, 218 pp.

D'Mello, P., 1997. Beach shacks fight for legitimacy. Herald, 12 September.

D'Souza, J.A. et al., 1988. The Regional plan for Goa, 2001 A.D. Town and Country Planning Department, Government of Goa, 108 pp.

Dume, A.R.S., 1986. The cultural history of Goa. Mapp Printers, Goa, 355 pp.

Frihy, O.E., Fanos, A.M., Khafagy, A.A. and Aesha, K.A.A., 1996. Human impacts on the coastal zone of Hurghada, northern Red Sea, Egypt. Geo-Mar Letts., 16:324-329.

Haq, B.U. and Milliman, J.D., 1996. Coastal vulnerability: hazards and strategies. In: J.D. Milliman and B.H. Haq (editors), Sea level rise and coastal subsidence, Kluwer Publishers, pp. 357-364.

Kale, V.S. and Rajaguru, S.N., 1983. Mid-Holocene fossil wood from Colva, Goa. Current Science, 32:778-779.

Kraus, N.C., 1988. Effects of sea walls on the beach. Jour. Coast Res., Sp. Issue, 4:1-28.

Lobo U., 1988. Environmental aspects of silica sand mining from coastal sand dunes. In: Earth Resources for Goa's Development, pp. 521-523.

Mascarenhas, A., 1990. Why sand dunes are needed. Herald, 21 December, p. 4.

Mascarenhas, A., 1996,a. The fate of sand dunes of Goa. In: Voices for the Oceans, G. Rajagopalan (editor), International Ocean Institute (India), p. 111.

Mascarenhas, A., 1996,b. Some observations on the coastal zone management plans of Goa (unpublished report), 25 pp.

Mascarenhas, A., Sawkar, K. and Chauhan, O.S., 1997. The coastal zone of Goa: then and now (Abstract 49). Seminar on Coastal Zone Environment and Management Mangalore University, February 12-14, pp. 50-52.

Mascarenhas, A., 1997. Coastal sand dunes ecosystems of Goa: significance, uses and anthropogenic impacts. Report under submission.

Mascarenhas, A. and Sawkar, K., 1998. Impact of tourism on the coastal ecosystems of Goa. Report submitted to the World Bank.

Miossec, A., 1988. The physical consequences of touristic development on the coastal zone as exemplified by the Atlantic coast of France between Gironde and Finistere. Ocean Shore. Manag., 11:303-318.

Narayan, R., 1997. Beaches under threat (editorial). Herald, 28 January, p. 4.

Nordstrom, K.F., 1990. The intrinsic value of depositional coastal landforms. Geog. Review, 80:68-81.

Nordstrom, K.F., 1994,a. Developed coasts. In: Coastal evolution: Late Quaternary shoreline dynamics, R.W.G. Carter and C.D. Woodroffe (Eds.), Cambridge University Press, London, pp. 477-510.

Nordstrom, K.F., 1994,b. Beaches and dunes of human-altered coasts. Prog. Phy. Geogr., 18:497-516.

Nordstrom, K.F. and Gares, P.A., 1990. Changes in the volume of coastal dunes in New Jersey, USA. Ocean and Shoreline Manag., 13:1-10.

Nordstrom, K.F. and Jackson,N.L., 1993. The role of wind direction in eolian transport on a narrow sandy beach. Earth Proc. Landforms, 18:675-685.

Nordstrom, K.F. and Mccluskey, J.M., 1985. The effects of houses and sand fences on the eolian sand budget at Fire Island, New York. Jour. Coast Res., 1:39-46.

Paskoff, R., 1994. Coastal zones - impact of managements on their evolution. Coll. Geogr., Paris, France, 256 pp.

Psuty, N.P., 1988. Balancing recreation and environmental system in a "natural area", Perdido Key, Florida, USA. Conservation and Recreation in the Coastal Zone. Fabbri, P. (ed.), Ocean Shore. Manag., 11:395-408.

Rosario, A., 1998. Industry and tourism take toll on Goa's coastline. The Navhind Times, 4 June, p. 10.

Sawkar, K., Noronha, L., Mascarenhas, A. and Chauhan, O.S., 1998. Tourism and environment: Issues of concern in the coastal zone of Goa. Special paper, Environment Series, World Bank (in press).

Sriram, K. and Prasad, K.N., 1988. Dune sands of Salcette coast - a study. In: Earth Resources for Goa's Development, pp. 617.

Subrahmanya, K.R., 1996. Tectonic, eustatic and isostatic changes along the Indian coast. In: J.D. Milliman and B.H. Haq (editors), Sea level rise and coastal subsidence, Kluwer Publishers, pp. 193-203.

Vogt, G., 1979. Adverse effects of recreation on sand dunes: a problem for coastal zone management. Coastal Zone Manag. Jour., 6:37-68.

Wilson, D., 1997. Strategies for sustainability: Lessons from Goa and Seychelles. In: Sustainable tourism? From Policies to Practice, M. Stabler (editor), CAB International Wallingford (in press).

13

SUSTAINABLE DEVELOPMENT OF COASTAL RESOURCES: BRACKISH WATER SHRIMP FARMING IN KARNATAKA

Ramachandra Bhatta

ABSTRACT

This Chapter analyses the institutions governing the use of the estuarine resources for shrimp farming in the Indian Coast of Karnataka State over the last two decades and the ecological and economic implications of this process. Three types of resource transactions, namely, market, administrative and technology have had their influence on the development of commercial shrimp farms. An increase in the international demand for shrimp and liberalisation in the import of technology of production have facilitated the allocation of brackish water area for shrimp farming. There is a growing evidence to show that intensive shrimp farm development has resulted in ecological degradation and threatened shrimp production itself. Even though the recent Supreme Court judgement ordered the closure of shrimp farms, its implementation may encounter stiff resistance from the shrimp industry. Therefore, an incentive-based tax system is suggested in order to force the shrimp producers to internalise their opportunity cost of production.

INTRODUCTION

Estuaries along the Indian coast of Karnataka have been historically developed into physical environment that sustains a multi-product renewable industry. In a typical estuary, a large brackish watershed is formed due to incursion of saltwater from the sea. Through a system of informal, collective farming arrangements, small groups of farmers in coastal areas have enhanced the productivity of estuaries by building and maintaining large system of levees to regulate the flow of saltwater. Reclaimed estuary lands are planted with salt-resistant rice varieties in the rainy season. In the dry season, as water salinity increases, the same watersheds become suitable for marine species, which migrate from the sea. During the last two decades, joint-farming groups have been increasingly producing shrimp in the dry season as a supplementary enterprise. However, the recent emergence of commercial aquaculture in the region has resulted in the breakdown of traditional resource management institution. Number of high-tech shrimp farms have taken over conventional croplands on lease, and have established commercial shrimp ponds. The recent government's land use policy also has served to hasten the process of estuary land conversion.

Background of the Study

The state of Karnataka is located on the western coast of South India, and has about 300 kms of coastline. The western Ghat, a series of mountains, runs almost parallel to the Arabian seashore, leaving a narrow belt of coastal land of only a few miles width. Several towns and villages along the shore are generally densely populated. Fishing constitutes an important livelihood for small farmers, land less labourers and fishermen. Fishes are an integral part of the human diet and the major source of animal protein for the lower and middle-income populations in coastal areas. The state's fisheries resource potential is estimated at 4,25,000 tonnes per annum over an expanse of 87,000 sq. km of the exclusive economic zone (Baktha, 1983). In addition to marine fisheries, coastal Karnataka also has a large potential for brackish water fisheries. Several major rivers originating from the Western Ghats and their tributaries flow through the coastal districts into the Arabian Sea forming a mosaic of more than 8,000 hectares of estuaries along the coast.

Traditionally, fishing had been undertaken through indigenous techniques with manually operated boats, and thus, can be characterised as a subsistence economic activity. In the last 20 years, fishing technology has undergone large-scale mechanisation with the help of government supported financial and educational programs. However, modern technology and capital have been accessible only by a small group of elite fishermen. A large class of traditional, small-scale fisherman either continued to operate with indigenous techniques or worked for modern fishing firms, and thus have not reaped the fruits of fishery mechanisation. Modern fishing vessels operated only in shore waters and overexploited the fishery. This has adversely affected fishing opportunity for small scale, indigenous fishermen. The traditional "rampani nets" which once accounted for 50 to 60 percent of the annual catch have almost disappeared (Bhatta 1996; Government of Karnataka,1996). These fishermen, therefore, have been forced to depend more on inland or estuarine species or seek employment with large fishing firms.

Even though fishery industry is an important economic activity in the coastal Karnataka, this region still is predominantly agrarian and about 60 percent of the total workers are dependent on agriculture and allied activities for their livelihood (Government of Karnataka, 1994). Food grains constituted a major portion (71%) of the croplands. Rice is the leading crop followed by peanuts and horticultural crops. A small portion of total cropland is devoted to high value crops such as banana, coconut and cashewnut. Since most of the croplands are within or near estuary and brackish water flood plains, fluctuating water salinity often influences crop productivity in the area. Therefore, management practices in the estuary, especially those regulating the water flow, could impact crop yields.

TRADITIONAL ESTUARY MANAGEMENT

Ecology and Institutions

Estuaries are one of the most biologically productive ecosystems in the world. People along the Karnataka coasts, through a co-operative management approach, have further enhanced the productivity of estuaries and created a sustainable environmental basis for multiple renewable resources (e.g., food, flsh, clean water, etc.). This section describes the ecological set-up of the traditional farming until the emergence of commercial aquaculture, and the socio-economic background of the estuary users.

The West Coast rivers of Karnataka turn into a vast expanse of brackish water areas near their mouths due to salt-water inundation from the sea. The tides travel several kilometres upstream flooding the low lying lands on either side of the rivers. Flooded lands along the natural estuary borders and river banks had been converted to permanent pond-like structures interspersed by natural streams (called Kodi) and man-made mud embankments with appropriate sluice gates (called Gandi) to regulate the flow of water. These embankments were fortified by planting mangrove trees. In addition to protecting the embankments from breaches, through entailment of roots, mangrove trees enrich the estuarine ecosystem. These reclaimed estuarine and river basin lands are called khar or gazani lands, and are put to variety of uses. Kodis, natural streams connecting gazanis and the nearest tidal rivers, provide a mechanism for regulating water in gazanis through out the year. Like gazanis, these streams also provide habitat for marine, brackish, and freshwater fauna depending on the season.

The estuaries and adjoining brackish water areas are subjected to great seasonal fluctuations in salinity. During the monsoon season (from June to August), heavy rains in the region and the catchment area of the rivers decrease water salinity. As a result, the estuaries become unsuitable for marine species. During this period, certain marine species migrate to the sea, and freshwater species colonise the estuaries. Starting in September, when the salinity of estuaries gradually increases, the marine species (including post-larval panaeid prawns, *penous mondon*) migrate to estuaries for feeding and growth. Thus, estuaries serve as a rich-fishing grounds round the year. They provide nutrient-rich substrates such as roots of mangrove trees and roots and debris of rice grown during the monsoon in gazanis. The mangrove vegetation along the embankment provides other consumptive and non-consumptive benefits such as sediment stabilisation, wildlife habitat, and a renewable supply of forest products, namely, boat construction wood, bark for tanning, fuel wood, fodder, fertiliser, and honey.

The gazani lands are used for cultivation of salt-resistant varieties of rice during the monsoon. The land is prepared by manual digging, with no animal- or power- drawn implements. The crops are sown in the first week of June just before the onset of monsoon. The seed must germinate before monsoon starts so that they do not get washed away. The only operation after sowing is harvesting. Manual operations are performed jointly by all those who own lands in a certain gazani. Rice cultivation in a traditional gazani is generally free of all chemicals, posing no toxic threat to marine or freshwater fauna colonising the rice fields or adjoining kodis.

During 1970s, in order to preserve the structural stability of the gazani lands and to sustain food production in the area, the state government undertook a massive project to build an extensive system of embankments that helped farmers to better regulate the water flow in case of heavy precipitation. They also promoted village transport and protected drinking water from the seepage of salt water from estuaries or gazani. Croplands adjacent to estuary rivers, which later have been converted to gazani lands, are privately owned properties. They can be classified as collectivised private resource (Kadekodi, 1997). In this case the individual farmers while retaining their property rights, pool their resources to collectivise and manage it as pool. Once again like the case of common property resources these lands have the characteristics of excludability, substractiveness and rules of sharing.

These lands get inundated with brackish water during the monsoon every year making it hard to identify and enforce the legal boundaries of individual properties. At best, inundated lands would rcmain as common capital "bad", giving no individual farmer any incentive to take unilateral actions for land improvement and cultivation. Therefore, land owners in each flood plain responded to this natural problem by forming informal, joint farming systems to collectively farm the brackish-water-covered (gazani) land. While individual farmers are still the legal owners of the land, this coalition was just an informal arrangement between farmers, an example of traditional transaction that helps member farmers to internalise the costs and benefits associated with the use of resource in question.

Under the leadership of a headman (Patgar) members contribute labour and other inputs to the collective farming efforts such as annual construction and maintenance of embankments and channels, planting, and harvesting. Output is distributed among members in proportion to the size of their land holdings in the joint farm. Most members are small farmers and grow crops mostly for subsistence. Members also collect fish and shrimp in gazani land.

Thus, landowners have been able to convert a private-property-turned-common--property 'bad" (inundated land) into a productive, more sustainable community resource. As indicated earlier, in order to protect the long-term sustainablility of gazani agriculture, the state government funded building of permanent embankments and terraces during 1970s. Prior to the establishment of these permanent structures, gazani farmers had to rebuild or fortify embankments every year. Embankments would last only the rice-growing season. Therefore, government investment on permanent embankment -an administrative transaction- was justified as a Pareto improvement policy. Pareto Improvement refers to a policy situation in which the sum of the gains, to whomsoever they were to accrue, exceeds the sum of the losses from that policy (Randall, 1987). The permanent embankments along with mangrove trees contributed to the long-term sustainability of the estuary-gazani resource system. In addition to providing small farmers with a long-term basis for subsistence farming, embankments contributed to the mangrove-related benefits to communities, promoted village transport infrastructure, and protected drinking water in surrounding villages from seepage of salt water.

However, the permanent embankment system turned out to be a perverse public intervention in traditional set-up of the estuaries. Around the same time when embankments were built, modern fish processing plants and fish wholesale contractors emerged in the region, creating new market demands for estuary fish and shrimp. The gazanis with permanent embankments in place could hold salt water round the year unlike traditional embankments, which would deteriorate after the monsoon and need re-strengthening every year. This enabled farmers to grow shrimp during the post-rice season. After the harvest of rice in November, farmers let tidal water into gazanis and stored salt water until the next April. Periodically, water was let out from gazanis to filter prawns that try to escape back to the ocean. One month before the onset of rains, gazanis are emptied of salt water to prepare the land for the next rice crop.

The emergence of shrimp filtration during the dry season was considered an economic boon for small, subsistence farmers who were seeking supplementary economic opportunities. Government agencies such as Brackish-water Fish Farmers Development Agency (BFDA) and Marine Products Export Development Agency (MPEDA) also promoted this cause by implementing extension and technical consultancy programs in order to train farm households in semi-intensive shrimp production technologies.

They also extended financial assistance with subsidy. The BFDA and MPFDA envisaged their technical and financial support programs for brackish water development under the assumptions that (Bhatta 1995):

- While adopting shrimp farming no significant diversion of the pattern suitable for rice crop would take place.
- Farmers will be able to realise the benefits of the new technology without any constraints.
- Income from other sources will remain unaffected and additional income will be generated with the adoption of brackish water aquaculture.

However, the course of this development seemed to have taken a wrong turn. Gazani farmers, who are financially weak and technically unskilled, opted to collectively lease their lands to fish contractors for undertaking year-round shrimp filtration activities and gave up their fishing rights. The efforts of BFDA and MPEDA were not enough to motivate farmers to undertake shrimp filtration on their own Furthermore, the overall marginal economic gain from leasing land for shrimp filtration did not appear to be as encouraging as originally anticipated. A survey of gazani farmers conducted during 1995 (Bhatta, 1996) revealed that the average rent from leasing the gazanis for shrimp filtration was around Rs.5300 per ha. Once the land was leased, farmers had to forego a net income from rice production of Rs.3000 per ha. The net incremental income from leasing was only Rs. 2300 per ha. Earning of this net income of Rs.2300 entailed other short-term opportunity costs such as

1. farmers had to forego their own fishing rights even in the kodis, which translated into a loss of Rs.750 per year household and
2. having lost the opportunity to grow rice in gazanis ,farmers had to buy more expensive rice-a staple food -from the open market at an additional cost of Rs.500 per tonne.

After accounting for economic losses in (a) and (b) above, the net marginal gain from leasing did not exceed Rs.1500 per ha. The only positive aspect of leasing was that farmers would get an assured total income of Rs.5300 per ha and avoid the risk of their crops being subjected to adverse weather and market conditions. Farmers could employ themselves in other income generating activities. The storage of salt water in gazanis through out most of the year increased soil salinity. With high levels of soil salinity, the ability of plants to absorb soil nutrients was severely restricted. In place where farmers had tried growing rice after years of shrimp filtration, crop yields were found to have declined from a normal average of 2 tonnes per ha to 0.5 tonnes per ha. However, the annual lease rent- a market indicator of demand for land - that farmers received from fish contractors did not account for the long-term productivity loss that the latter sustained. This supports the view that the gazani land allocation for shrimp filtration was not done in an efficient manner.

Gazani land leasing for shrimp filtration also affected another group of traditional users of estuarine system. Ambigas (land less fishermen) have traditionally had free access to fish in smaller water canals and kodis. When gazanis were leased out, Ambigas were disallowed from fishing in the kodis. The High Court of Karnataka upheld their fishing rights. Nevertheless, the lack of enforcement of the High Court order and the political and financial power of fish contractors made it almost impossible for Ambigas to claim and enforce their traditional rights. Tens of hundreds of fishermen were left in great economic distress as they had very limited skill and education to seek alternative gainful employment.

In summary, a host of economic and political factors contributed partial decline of traditional estuary-gazani management regimes, even before the emergence of commercial shrimp farms in recent years. First, farmers had formed an informal joint farming association to handle the natural flood problem. Estuary resources were exploited in an ecologically sustainable way just to meet their subsistence requirement. Later with the direct government intervention into the traditional estuary set-up, even though for a good cause, land owners and commercial fish contractors sought opportunities to turn traditional gazanis into profit-making assets by producing high-value shrimp. In the long run, wherever the lease period had expired or the external shrimp contractors had pulled out of gazanis for economic or regulatory reasons, farmers were stuck with lower un-productive gazani lands.

EMERGENCE OF COMMERCIAL SHRIMP FARMS

Shrimp culture in India in general, and in Karnataka in particular, took a new trend in early 1990s. With the rising foreign and domestic demands for shrimp, private corporations saw a golden opportunity in the estuary resources. A wave of modern, scientific shrimp farms swept coastal Karnataka. Both multinational and domestic corporations took over gazani lands on long-term leases and established high-tech, capital-intensive shrimp ponds.

The two districts together consist of over 4000 hectares of brackish water area suitable for prawn farming. Out of this roughly 2500 hectares are classified as khar-lands. Of the total khar-lands about 700 hectares are still under traditional prawn farming, while the rest have been developed for scientific prawn farming, either extensive or semi-intensive. Semi-intensive culture of *penaeus monodon* has been practiced at a density of $10/m^2$ and in the remaining areas, extensive culture has been practiced at a density of $4/m^2$. The average yield from a semi-intensive culture is around 2000 kg per hectare, while under extensive method it is around 750 kg per hectare.

The brooders of *penacus monodon* are obtained by hatcheries from middlemen who in turn purchase them from the fishermen from the East Coast. The fishermen collect brooders in an open access trawl fishing. At present each brooder prawn would cost around Rs. 250-300. The middlemen in turn transport the brooders to shrimp hatcheries. There are 11 shrimp hatcheries in Karnataka with a capacity of around 500 million seed production per year. These hatcheries supply seeds up to Bangladesh and Sri Lanka, apart from local buyers from Kerala, Karnataka and Goa. The total seed demand in Karnataka is estimated to be around 100 million seeds per year with a stocking density of 10,000 seeds per hectare.

Nursery Ponds

The areas of these ponds are about 0.1 ha with a water depth of about 80 cm. The post larvae are stocked at density of one million per hectare. The survival in the nursery ponds is about 70% and the post larvae should reach an average weight of about 1 gm after 50 days. The post larvae are fed a variety of food such as minced clam or animal or plant waste or manufactured feed. As the exact dietary requirements of the shrimp is not as well understood as it is for salmonid fish, the formulation of these feeds cannot be as exact. Often ,chicken pellets are used for the shrimp feed. It is probable that much of the feed is not consumed directly by the shrimp but acts as a fertilizer stimulating the growth of the bottom mat of algae on which the shrimp graze. The feeding rate varies with the stage of development of the shrimp and is related to their body weight. For the first 7 days 200% of the body weight is fed per day, then 50% for the next 14 days and 25% for the last 21 days in the nursery ponds.

Growout ponds

The size of these ponds varies considerably from well below 1 ha to many hectares in area. Many farmers consider that an area about 2 ha is the best size. such ponds have a sloping bottom and a average depth of 1 m. The shrimp from the nursery ponds are stocked at a density of 1,00,000 per hectare. Depending on water temperature and feed, the shrimp can reach a marketable weight of 25g in 4-6 months. The survival during this period is about 90%, the actual figure depending on the skill of the farmer. The shrimp are fed a variety of feed, depending on local availability and the sophistication of the farming procedures used. These feeds include clam meat, animal and plant wastes, formulated pellets and chicken pellets. Again, as in the nursery ponds, a significant amount of the food is probably acting as a fertilizer for the production of the algae mat. When clam meat is used as a feed an overall conversion of 10:1 is often achieved, with pellets, conversion ratios of 4:1 and some times 2:1 are obtained.

Pond Management and Production Levels

Shrimp farming in many areas of the world has recently led to severe problems of environmental degradation. The control of water quality is therefore a key factor for regulating success in shrimp farming. Salinity, temperature, oxygen level and hydrogen sulphide level are the most important parameters for the management of the pond water. Regulation of these factors is mainly achieved by control of water. Regulation of these factors is mainly achieved by control of water exchange, either by the use of tidal flow or by pumping. Oxygen levels are often maintained by the use of mechanical agitators to aerate the water. Hydrogen sulphide levels are controlled by the use of iron oxide that is added at the rate of 1 kg/m.sq. of pond bottom per day. Water exchange is however, of the utmost importance and intensive feeding systems require an exchange rate of 10% day which can be reduced to 5% in less intensive systems. It has been calculated that, if water exchange rates of 33% day are possible, very high production levels for shrimp, in the order of 5000 kg/ha/year can be achieved.

The production levels vary enormously from farm to farm depending on the farm management and the type of farming practice, the levels varying from a few hundred kg up to in excess of 3000/ha/year (Tables 13.1 and 13.2).

Harvesting can often take place about 120-150 days after stocking when the shrimp sizes will be about 22-33 gms weight. Shrimp are frequently marketed by their size, calculated as the number per pound (lb), either whole or as tails. In intensive Systems counts of 3040 per pound of head less shrimp can be obtained. Shrimp seeds that were artificially grown in hatcheries replaced the naturally grown post-larval shrimp immigrating from thc sea.

Legislative Response: A Catalyst for Shrimp Farm Development

The recent emergence of capital intensive shrimp farming and decline in traditional agriculture can partially be attributed to evolution of land reform regulation in the state. For instance, a 1961 land reform legislation of Karnataka aimed at maintaining more equity in asset distribution in rural areas(Government of Karnataka, 1961). With the intention of preventing agricultural lands from being diverted to industrial uses, the legislation specially outlawed leasing of farmland for any purpose other than six traditional agriculture activities: horticulture, crop production, grass and garden produce, dairy farming, poultry farming, and breeding livestock. Also, a ceiling on the size of land holdings per person of 10 units (an equivalent of 87.45 ha of dry land) was placed. These provisions prevented non-agricultural capitalists from acquiring and consolidating agricultural lands into large corporate farms. The legislation's goal was to discourage absentee land-lordism in agriculture and disparity in rural wealth distribution.

Table 13.1: Production Economics per Crop per Hectare (In Rupees)

1. Lease cost per crop	9000
2. Main workers for 4 months@Rs. 1000/per month	4000
3. Casual labour	
(a) Pond preparation for 5 days @ Rs. *50/day*	250
(b) harvesting	1000
(c) repairs	1000
4. Net/ Screen (4 in 1ha)	1000
5. Cat walk & cheek ways	1000
6. Planks for the gate 1/2 inch thick 18 Sq.ft. @ Rs. 20/Sq.ft.	400
7. Diesel cost Rs.9/litre x 5 litres/day x 20 days/ month + Oil	4000
8. Lime 400 kgs for pond preparation @ Rs.3/kg	1200
9. Lime during the crop 1 ton @ Rs. 2/kg	2000
10. Fertilisers 30-50 kgs before stocking @ Rs. 12/kg x 40kgs	500
11.Ammonium Sulphate 40 kgs mixed with burnt lime for killing fish	400
12.Seeds 6 Pieces/m.sq. x Rs. 50 x 10,000 m.sq.	30,000
13.Feeds (60,000 seeds x 0.80 survival x 25 gms size = 1.2 tons x 0.9 (FCR) 1 ton feed x Rs. *45/kg)*	45,000
Total operational cost	1,08,000
14. Income 60,000 seeds x 0.80 survival x 25 gms = 1.2 tons x 0.9 (FCR) = 1.2 ton @ Rs. 250/kgs	2,88,000
Gross Profit	1,80,000

Table 13.2: Comparison of the Profitability (June 1997)

Item	Goa	Kundapur
Pond area (WSA) (ha)	0.8	*0.5*
Stocking date	31-1-97	11-3-97
Harvesting date	6-6-97	8-6-97
Culture days (days)	126	87
Initial stock pieces	40,000	20,000
Stocking density (pieces/ m. Sq.)	3	4
Total Shrimp harvest (kg)	1135	544
Harvested Size (gms/piece)	33	32
Survival Rate	86%	*85.5%*
Total feed consumed (kgs)	975	473.3
FCR	0.86	0.87
Selling Price (Rs/kg)	320	250
Income (Rs.)	3,63, 200.00	1,64,396.00
Production cost		
Seed price (Rs.)	0.60	0.40
Total seed cost (Rs.)	24,000	8000.00
Feed price (Rs.)	*45.00*	45.00
Total feed cost (Rs.)	43875.00	21298.00
Profit		
A Income (Rs.)	3,63,200.00	1,64,396.00
B Seed cost (Rs.)	24,000.00	8000.00
C Feed Cost(Rs.)	43,875.00	21298.00
Gross Profit (A-(B+C)) (Rs.)	2,95,325.00	1,35,098.00
Seed cost/kg of shrimp (Rs.)	21.14	14.71
Feed Cost/kg of shrimp(Rs.)	38.66	39.15
Productivity/ha/crop(kg)	1419	1088
Average daily gain (kg)	0.262	0.367
Salinity range (1)pt)	15-33 ppt.	25-33 ppt.
Feeding method	Boat feeding	Broad casting
Feed brand	Bintang	Bintang

However, the 1995 amendments to this law made land leasing for aquaculture an exception (Government of Karnataka, 1995*)*. Aquaculture was accorded the same status of the above six agricultural activities. Thus, croplands now could be leased out for aquaculture for a period not exceeding 20 years. Also, the purchase of croplands was made easier by raising the land ceiling to 40 units per person (or 200 units for a standard family of 5 members, equivalent of 427.53 hectares per family). The state government had power to relax this limit further under special circumstances. The 1995 amendments also relaxed the category of those who were allowed under 1974 Act to acquire agricultural land. Households earning more than Rs. 50,000 from non-agricultural sources were previously prevented from acquiring croplands. This limit was raised to Rs.2,00,000. These land policy revisions together empower prospective shrimp growers with ability to operate larger shrimp ponds resulting in lower pond construction and operational costs. As a result, more than 60 percent of the area under shrimp ponds is now owned by large industrial enterprises(Government of Karnataka, 1996).

The above legislative reforms have served as a catalyst for shrimp pond development by subjecting coastal land use decisions to more open market competition. In this process, one can see an evolutionary change in the institutions governing land uses: a change from traditional transaction (joint farming by farmers) to market transaction (leasing and purchasing), facilitated by administrative transaction (legislative reform). Therefore, the decline of traditional estuary management regime can be viewed as an end result of an institutional-evolution process. The crux of this evolutionary process is to change the structure of 'land development rights" that different users have had. The development rights were originally held by gazani and other cropland owners. Landowners themselves started giving up this right for obvious economic and social reasons, i.e., by way of leasing land first to shrimp filtration and later to modern shrimp farming. Because of the tremendous economic opportunity associated with the rising foreign shrimp market, private industries (seeking economic rent), and the state government (seeking more tax revenue and yielding to the political pressure from industries) claimed a larger share in the development right. This change in the right structure was made possible using both market power (high land lease rent) and administrative power (legislative reforms).

Disease and Declining Profitability of Shrimp Production

Since 1994 the white spot disease-affecting shrimps have been playing havoc in India. Prior to this the disease has had devastating effects on the shrimp culture industry in China, Taiwan, Thailand and other shrimp producing countries. As of this day no scientifically validated management technology has been developed to control this disease in the region. During 1994-95 and 1995-96 growing season, viral disease had infected more than 1000 hectares of shrimp ponds. The disease has led to stunted growth and death causing enormous loss to firms. It was estimated that an input loss of (due to the death of juvenile shrimps) Rs. 10 million per year was incurred. Also, an output loss of Rs. 25 million was estimated for the same period. Some shrimp producers have tried to grow alternative fish species with little success. In addition to this production risk, shrimp producers also could be subjected to market risks. International markets are becoming competitive with increased production in almost every shrimp-producing country in the 1990s. If this trend continues, market prices could decline in the future affecting the economic viability of the shrimp industry . In the presence of both production and market risks, the flow of economic benefits from shrimp production cannot be sustained in the future.

Environmentalism and Judicial Response

What government cannot achieve sometimes can be accomplished by people- if they act together. Environmentalists in India assumed an active role in influencing ultimately who owns the development rights to estuary systems. However, in this case they did not resort to the usual picketing or public protest. Instead, they challenged the rights of shrimp producers by filing a public interest suit against them in the Indian Supreme Court. In response to this law suit, the Supreme Court ordered the closure of all the shrimp ponds operating within 500 meters of high tide line (TET, 1996). This ruling is basically a reinforcement of an earlier regulation of the Government of India's Ministry of Environment and Forest that had disallowed any form of development activities in Coastal Regulation Zones. Neither central government nor state governments had enforced that regulation on aquaculture. This laxity in enforcement forced environmentalists to seek the intervention of the judicial system.

As per the order of the Supreme Court ruled in December 1996, the affected shrimp farms must have been closed down by the end of March 1997. The order allows the continuation of traditional shrimp production (e.g., shrimp filtration) activities by fishermen. The court also ordered that an authority be constituted for overseeing the closing down of shrimp farms. This authority is required to collect compensation from shrimp farms for the damage they caused in terms of soil erosion and depletion. Shrimp farms also are required to pay compensation to people and workmen who would loose incomes and jobs once the farms are closed.

Because of the irreversible nature of shrimp pond development, would shrimp producers be made accountable to farmers for the croplands lost in perpetuity? Once those ponds are closed, landowners will neither receive lease rent nor crop incomes.

Would the proposed authority collect damage compensation from shrimp producers toward the groundwater contamination? Also, the court order does not affect ponds that are beyond 500 metro landward side from the high tide line. These ponds also are causes of concern for stakeholders and residents in that area. Therefore, the court order cannot be viewed as an ultimate panacea for all the problems.

Discussion and Policy Conclusions

The estuary resources of Karnataka have undergone rapid changes in its utilisation with the advent of coastal aquaculture. During the early 60s' the gazani land were used for growing salt resistant paddy varieties with the average yield of 2-3 tonnes per crop per hectare. During the late 70's and 80's paddy-cum-prawn filtration method was introduced into the system. Though it did not lead to substantial increase in the average income of the farmers there was enough evidence to show that a new class of entrepreneurs had emerged making a good profit from taking gazani land on lease basis and stocking shrimp using the traditional filtration method. Then came the imported scientific commercial shrimp farming technology.

Many Indian and multinational companies invested huge capital on hatchery, farming, feed production, and processing and export trade. This was further facilitated by the amendments to Karnataka Land Reforms Act in 1995, which included the aquaculture within the definition of agriculture and allowed the conversion of coastal estuary land for shrimp farming. The new legislation also paved the way for commercial shrimp farming by legalising the leasing out of farmland, relaxation of investment limit by outside capitalists in agriculture. However, the new technology was unsustainable as was revealed by the outbreak of viral disease to cultured shrimp and the socioeconomic impact in terms of reduction of the drinking water supply, reduced yield of agricultural crops in the adjacent land and labour displacement.

Thus, apart from several far reaching effects on the ecologically fragile coastal ecosystem the new technology has led to a loss of several millions of rupees in terms of capital investment and loss of alternative crops. Now that the Supreme Court of India has ordered the closure of all the shrimp farms a big vacuum has been created without any direction for proper use of these resources. The owners of the gazani land have lost their traditional income from paddy cultivation as well as lease income. Now it is neither possible to convert these shrimp farm into paddy fields nor any suitable alternative is available for culture.

The Supreme Court's judgement to close down the commercial shrimp farms is an indication of the damages caused to the coastal ecosystem. Hence the demise of the large-scale shrimp farms need not be mourned. However, since the land can not be reclaimed back to paddy in what form the land could be utilised in a sustainable manner.

The question of ecological balance revolves around the issues like whether short term superficial gains should loose sight of the permanent damage that they may cause. Loss of paddy crop for apparently profitable prawn farming and the present predicament of the owners of these lands should be seen as a short sighted penalized both by man and nature. Clearly low income from paddy cultivation have prompted the farmers to lease out their land for shrimp farming which led to degradation of the coastal ecosystem. Are their any other means of restoring the diversity and productivity of these resources in a manner more consistent with the resource needs of the local communities? Thc traditional prawn filtration method suggest that the farmers can successfully develop their resources sustainably and collectively for community benefit. However, the Government of Karnataka may not think so judged by the amendments to land reforms legislation to invite large-scale capital for shrimp farming and tourism projects.

The government incentives and subsidies now fueling the urban investment can be re-channelled towards village organisations to catalyse such an alternative coastal resource development strategy. The judgement of the apex court should serve as a lesson on the importance of the sustainable coastal resource development.

REFERENCES

Baktha, N.P.(1983): Development of infrastructure facilities for the optimal exploitation of the Exclusive Economic Zone. in *Fisheijes Development in India,* ed. U.K. Stivastava and D.M. Reddy, pp 313-326. New Delhi: Concept Publishing Company.

Bhatta, R. (1995): The impact of new agricultural policy on brackishwater aquaculture development in Kamataka. Presented at the New Agricultural Policy Workshop,Mangalore, India, August 1995: lOpp.

Bhatta, R.(1996): Socioeconomic aspects of brackish water aquaculture in Kaniataka. Report submitted to the Developmental Education and Communication Unit, Indian Space Research Institution, Ahmedabad, India: 66pp.

Government of Karnataka(1961): *Karnataka Land Reforms Act, 1961.* Government of Karnataka, Bangalore,lndia.

Government of Karnataka (1995): *Karanataka Land Reforms (Amendment) Act,1995.* Government of Kamataka, Bangalore, India.

Government of Kamataka (1996): *The Report of the committee on Coastal Shrimp Farming.* Department of Fisheries, Government of Kamataka, Bangalore,India.

Kadekodi Gopal (1997): The common pool resources: An institutional movement from open access to common property resources, *Ener~' Resources,* forthcoming.

Randall A (1987): Resource Economics. New York John Wiley & Sons 434 p.

TET (1996): SC order demolition of prawn farms to bail out fishermen. The *Economic Times* December 12,1996:1-26.

14

IMPACT OF SOCIO-ECONOMIC DEVELOPMENT ON THE MANGROVE ECOSYSTEM OF COCHIN IN KERALA

Vijaya Subramanian

INTRODUCTION

In recent years, people who are genuinely interested in wildlife protection and nature conservation have expressed serious concern about unplanned development activities initiated along the coastal zone of our country which come into conflict with preserving threatened ecosystems such as the mangroves.

The mangroves are salt tolerant inter tidal communities of trees and shrubs thriving on loose muddy soils and substratum created by tidal action and river run off especially in the estuarine regions of major rivers along our coastline. Sunderbans, for example, is the largest single block of mangrove forest in the South East Asian Region occupying about 4000 square kilometres. Mangrove areas of lesser extent of 5000-15000 hectares are encountered elsewhere in the estuaries of Mahanadi, Godavari, Krishna, Cauvery and along smaller river systems and salt water creeks along the west coast.

Cochin harbour area which is backed by an extensive estuarine system known as Cochin backwaters has many pockets of mangrove habitats often obscured by the dense fringes of coconut plantations. These habitats have the same species diversity which one finds in any other mangrove ecosystem. In Cochin the mangroves confer many direct and indirect benefits such as protection of shoreline, prevention of coastal erosion, dampening of strong winds and cyclones, nutrient recycling, maintenance of estuarine water quality and as shelter and nursery grounds to a wide range of aquatic species of commercial importance. Besides these, the mangroves, as a forest resource, has been yielding timber, fuel wood, charcoal, honey and other products of economic importance to local inhabitants.

The mangroves, with their swampy soils, clusters of breathing roots, prop roots and tangle of trees and twiners have often been considered as a waste land or a dangerous place inhabited by harmful wild animals and thus their intrinsic values have often been overlooked when any development activity is undertaken due to compulsions of population pressure and the consequent socio-economic development needs. A balanced approach which would satisfy both development and conservation needs is stressed now a days thanks to greater knowledge, awareness and active judiciary.

With this background, the impact of socio-economic development activities on the fragile mangroves of Cochin is reviewed and evaluated based on the existing material and literature.

COCHIN AND ITS ENVIRONS

The harbour city of Cochin is situated around 10°N latitude in Kerala coast, facing the Arabian sea. The harbour has a vast hinterland supported by the foot hills of the Western Ghats. The coastal plains and an extensive estuarine system is formed by the Vembanad Lake and Cochin backwaters. More than two dozen small rivers apart from the larger Periyar river empty the fresh water into this lake and the tidal waters from the sea enter the Cochin bar mouths and penetrate deep into the system, thus maintaining an estuarine complex. In the vicinity of Cochin harbour itself there are about 6-7 islands formed by narrow coastal spits and silt deposition or accretion. The Vypeen island runs as a narrow strip from the Cochin bar mouth to Munambam, about 40 km North. The other islands are of about 1-5 km length with narrow width. All these islands are now populated with various degrees of agricultural fisheries and aquaculture activities.

The present day "Cochin" is actually a merger of two towns namely Ernakulam on the eastern side of the back water and the original Cochin on the western side including the harbour. The Willingdon island which provides the berth to the ships and house the Port Trust is a man-made island formed by reclamation of the back waters. The marine drive in Ernakulam stretching to about 2 kms is also through reclamation.

In many locations along the Vypeen island, Fort Cochin, Kannamally, reclaimed islets and along the salt water creeks and canals in the mainland, mangrove patches are encountered. Further, deep into the estuary, also there are many locations having mangrove formation as far as 10-15 kms north and south of Cochin.

Structure and Composition of the Mangroves

The mangroves that we come across in and around Cochin comprise of woody species such as Avicennia, Rhizophora, Bruguiera, Excoecaria etc. and shrubby forms like Acanthus, clerodendoron, Aegicaras, etc. and also a mangrove fern Aerostichum aureum. Towards the landward fringes, mangrove associate ferns like Thespesia populnea, Hibiscus tiliaceous, Terminalia catappa, Pandarus sp., Artocarpus spp are often encountered. Towards the beach side of lagoons, strand vegetation such as Ipomoea pescaprae, Acanthus sp., Panicum spp. are common.

The number of species in such island on mangrove habitat varies from location to location. For example, in "Mangalvanam", a mangrove habitat under conservation, undisturbed tree forms of Avicennia dominate in numbers with occasional patches of Rhizophora or Acanthus, thus making up the total number of species to 3 or 4. Whereas, in the southern tip of Vypeen island, where the mangroves are growing in sea accreted areas, one can count about 11 species.

Tree density per 100 sq.m is usually 150-200 in respect of older trees of diameter above 5 cm and 250-300 in the case of younger stems above 2.5 cm. The canopy heights in well preserved pockets range from 8-12 m.

ECONOMIC AND ECOLOGICAL IMPORTANCE

As already stated, mangroves confer many direct and indirect benefits to the coastal population. The annual turnover of wood biomass is utilized for firewood, charcoal, fishing poles, piles etc. The foliage is also utilized as cattle feed. The most important utility is the litter production in mangrove areas which is estimated at 5-10 t/ha. This litter upon degradation becomes detritus which is utilized by a number of species of prawns and fishes. Certain species spend part of their life cycle in the mangrove dominated estuary and later migrate into the sea where they form an important resource for coastal fisheries. Various forms of shrimps are example of this category.

In Cochin estuarine system, the mangrove based water bodies are visited by more than 8 species of prawns, 12 species of estuarine fishes and over 50 species of marine species near the harbour area. Since the mangrove areas also provide good shelter for juvenile forms, these areas are exploited by traditional prawn and fish farmers for collecting seeds of prawns and fishes for rearing them in farms.

The Cochin mangrove area is well know for "Chemmen Kettu" (Trapping and holding of seeds in farms influenced by tidal water) and also the Paddy-cum-Prawn culture. "Pokkali" is the rice variety that is cultivated alternately with prawn culture in the same field with the impetus given to coastal acquaculture many seasonal and perennial ponds/farms have come up in many locations in the islands and near shore areas. A number of people are engaged in these activities for their livelihood.

Taking advantage of daily migration into and out of the estuary by a number of species of fishes, there is a regular sustenance fishery going on in the estuary. The use of "stake nets" and "Chinese Dip Nets" are widely prevalent in the area, apart from the use of gill nets and cast nets. About 8 to 10 thousand tonnes of fishes are harvested annually from these backwaters. The mangrove detritus is one of the source of food to sustain this production. In the coastal agriculture operation the fish farmers are able to get a production around 500-600 kg/ha/season.

The detritus production from mangrove and other sources, the fish and prawn seed biomass produces along with primary producers like plankton find their way throughout the Cochin bar mouth to the coastal waters and enrich the coastal fish production. The fish production in the coastal areas of Kerala contribute significantly to the marine fish production of the country. The contribution of pelagic species such as oil sardine mackerel and small tunas and the penacial prawns is very well known. The mangroves in this area play a vital role in the food web of mangrove dependent fish and prawns resources.

From the ecological point of view the mangrove areas, especially the preserved ones, serve as sanctuaries to a number of aquatic birds such as little black Cormorants, Ibises, Painted Storks, Darter, Egrets and Herons. Many species of clams, oysters, gastropods and crabs are resident in mangrove areas. The species diversity in a protected mangrove habitat is well known.

Socio-Economic Developments and its Impact

Having outlined the positive side of the mangrove areas and their economic importance, the impact of development activities in this area on the mangrove is reviewed.

A conservative estimate indicates that the total extent of mangrove areas in the Cochin Backwaters and Vembanad lake was around 70,000 ha. This area got progressively reduced as the mangrove areas were converted for coconut plantations, paddy cultivation, traditional pond culture, reclamation and other development activities.

According to UNESCO Expert Martha Vannucci about 61 km^2 of "Kari" land with dark peaty soil having high proportion of carbonaceous wood contributed by mangroves have been converted for agriculture in the Southern region, namely, Kuttanad area. An equal area has been converted for paddy-cum-prawn culture in the middle and northern sectors of the Cochin estuary (Vypeen, Narakkal, Edvanacad, Cherai, Parur and other areas).

Reclamation of back waters, once bordered with mangroves have gone on a steady pace unmindful of the destruction or damage to the ecosystem. Some examples are as follows:

- creation of the Willingdon island for Cochin harbour and port development (364 ha)
- Creation of Fisheries harbour (11ha)
- Conversion of Vallarpadam and Ramanthuruthu island for port development (140 ha)
- Southern extension to Willingdon island
- Naval base quarter etc. (141 ha)
- Marine drive project of GDCA (24 ha)
- Construction of shipyard
- Construction of super oil tanker berth etc.

Various forms of threats to the mangrove ecosystem is continuing as a result of population pressure and growing need of land, for housing, urban and industrial development. Wetlands and mangrove swamps have been filled up in the recent past for housing projects such as Giri Nagar, Panampilly Nagar, Jawahar Nagar, Kumaran Asan

Nagar etc. The construction of Cochin bye pass highway from Edapally to Kumbalam for a distance of about 16 km has wiped out many pockets of mangrove swamps.

An undistributed patch of mangrove area extending to about 3 km on the Western side of Vypeen island (southern tip) is threatened with destruction for the purpose of development of a gas based industry and a power plant. In addition, about 500 ha are threatened for conversion for the purpose of linking the various islands with road bridges.

Population Increase and its Consequences

The Cochin city is fast growing as a major metropolis with people from other districts and States migrating to it for the purpose of employment, business and education. The Cochin Naval base, Port Trust, Shipyard, Customs, Central excise, NPOL, CSIR and ICAR Institutes, MPEDA, Telecom and Railways are some of the Central Government Institutions which were established during the last 25-30 years and which brought tremendous pressure on the available land area around the backwaters, for the purpose of their offices and residential quarters. The State Government has created about 40 establishments in and around Cochin.

With the establishment of the Fisheries Harbour, mechanized fishing operations have increased ten fold compared to what it was in the early sixties. Some relevant statistics are given in Table 14.1.

As the mangroves and fisheries are closely related let us examine certain facts. Along the coastal strips of Vypeen island the population density of fishermen and agricultural workers is one of the highest in the country at about 1500-1600 per km^2.

The villages are contigerous stretching to about 40 km. The fishermen and other poor people put up their dwelling units by encroaching on the mangrove areas and beaches. To carry on their fishing activities, a number of smaller boat jetties and berthing places have been developed. The mechanized boats and motorized boats traversing through the canals and creeks pollute the mangrove waters as a result of oil spills, kerosene discharges and also due to washing of nets, cleaning of fish etc. For the purpose of fish and prawn export there are a number of processing plants from where waste water and other organic wastes are discharged into the backwaters. The discharge of sewage and disposal of garbage and solid waste from the municipalities add to the problem of pollution. As the mangrove waters get increasingly polluted the population of fish and prawn seeds get reduced.

Table 14.1: Some Socio-economic Parameters of Cochin's Ecosystem

Parameter	Value
Total population in Ernakulam District	2,797,779
Average density per km^2	1162
Growth rate during the period 1981-91	10.36%
Total Fisherman population in the Dist.	67,787
Total Fisherman population in the Dist. (Males)	21,813
Total Fisherman population in the Dist. (Females)	21,335
Total Fisherman population in the Dist. (Children)	24,639
Number of Coastal villages	21
Number of mechanised boats	1241
Number of motorised country crafts	465
Number of ordinary country crafts	1740
Average fish catch from the district (tonnes per annum)	95,567
Average shrimp catch (tonnes per annum)	10,022

Source: Govt. of Kerala , Kerala Fisheries- Facts and Figures, Dept of Fishery(1992).

The fishermen having settled close to the sea shore raised the problem of sea erosion and wave action threatening their dwellings. In order to protect these dwellings, the govt. of Kerala has constructed granite walls and "groins" along this cost thus interfering with the inter tidal zone and thereby the mangroves. At several locations in the back water the islets have been bunded with granite walls for developing water ways and prevent silting.

Elsewhere in the outskirts of this region major industry such as FACT, Hindustan Insecticides, Indian Rare Earths, Travancore Rayons etc. empty their effluents into the Periyar river and these harmful chemicals find their way into the back waters and also the mangrove areas. Leaching of pesticides and fertilizers from adjoining agricultural fields are also other sources of pollution.

The construction of a salt water barriage at Thannermukom (near Mangrove area W-2) and the construction of a flood control spill way at Thottapally have brought about changes in abiotic and biotic factors of the estuary. Such obstructions interfere with the free in-flow and out-flow of tidal waters and thus alters the estuarine quality of the water bodies such as salinity and pH and consequently the movement of larvae and juveniles of fishes and prawns. Constructions of bunds and barriages across the estuary, reclamation within the back waters for developing port facilities, extension of airport, road bridges connecting the islands are some of the examples of man made changes that effect ecological balance.

CONCLUDING REMARKS

Human beings should realize that they are the part and parcel of nature surrounding them. They are one of the entities of a larger eco-system or a biosphere. If human population is alone allowed to increase without control especially in a limited area such as the coastal zone, one or the other plant or animal community gets depleted or degenerated.

As population increases, there is more and more demand for food, shelter and employment. Politicians, planners and administrators rush for grabbing the available land area for initiating development projects to meet the above mentioned needs. When Scientists and Conservationists raise objections, these are often ignored. When the question of benefits for "man" or survival of "mangrove" is debated in legislature or council, "man" gets preference and "mangrove" is sacrificed. This is not the right approach for a sustainable development.

It is to be noted that about 20-25 years back, Ernakulam and Cochin were small and peaceful towns with lots of avenue trees and other vegetation. There were a number of swamps and wet lands from where frogs used to come out in the night and frog exporters used to catch them. Traditionally, Malayalees loved to have an independent house with fruit bearing trees in the compound. One could see the stars in the sky during the non-monsoon months. But now all these have changed. The landscape is getting transformed with monstrous high rise buildings getting out of the palm fringed shores. The calm backwaters are churned by oil tankers and hundred of fishing boats. There is more and more demand for fish, rice and other commodities.

Some experts are skeptical whether we can meet all the demands of a growing population. Eminent scientists and agriculturist have drawn attention to certain facts which are mentioned below:

- Increasing population leads to increased demand for food and reduced per capita availability of arable land and water for irrigation,

- Increased purchasing power and increased urbanization lead to higher per capita food requirement due to consumption of animal products,
- Marine fish production has been stagnating since 1990, and
- There is increasing damage to ecological foundations of agriculture such as land, forest, water, biodiversity and the atmosphere and there are distinct possibilities of adverse changes in climate and sea level.

What is needed is a balanced approach to development activities and conservation measures. The carrying capacity of watershed and land area should be properly assessed and scientific land and water use planning should be undertaken so that need based development activities ensure a sustainable ecological basis.

REFERENCES

Govt.of India, (1987): Mangroves in India- A status report, Ministry of Environment and Forests, Government of India.

Govt. of Kerala, (1992): Kerala Fisheries - Facts and Figures, Department of Fishery, Kerala State.

Rajagopalan M.S. (1992): Studies on some aspects on some, mangrove ecosystem in a tropical estuary.-Ph.D. Thesis, Cochin University (pp 154)

Gopalan, et al. (1987): The sinking backwaters of Kerala, JMAR Biological association of India-25; 131; 141.

Vannucci (1984): Conversion of mangrove areas for other uses, workshop on human population, mangrove resources, and human under stress BOGOR, Indonesia.

UNESCO (1979): Mangrove Ecosystem - human uses and management implication, UNESCO report on marine science no.9 (pp 46).

Pillai K.K. (1990): Water quality studies in Cochin estuary system paper presented, seminar on water quality status of Kerala, CWRDM, Calicut.

15

COASTAL PROCESSES IN TAMIL NADU AND KERALA: ENVIRONMENTAL AND SOCIO-ECONOMIC ASPECTS

J. S. Mani

INTRODUCTION

Vast coastal front of a country signifies unlimited potential by way of both living and non-living resources. India with nine coastal states and a couple of union territories possesses precious water spread extending for about two million square kilometres which is designated as Exclusive Economic Zone. Though it may not be a simple task for a developing country like India to harvest living and non-living resources in the deep ocean region (due to limited technological capabilities), the nearshore region due to easy access and traditional practices, has been subjected to over exploitation in the recent past. In addition, the fragile coastal zone is compelled to accommodate the surge of population migrating from inland region. The combined effect of population influx and over exploitation has threatened the coastal zone and associated eco-system which, in due course of time, will disturb the coastal equilibrium. Of late, construction activities along the coastal zone by private and government agencies have contributed to the coastal inequilibrium mainly because of prevention of natural sediment movement by waves.

Under this situation the coastal zone frantically tries to readjust so that a state of equilibrium between the various forces can be maintained.

The coastal region is so delicate that it takes time to readjust and often emergence of coastal structures at frequent interval perplex the coastal region eventually resulting in unprecedented coastal deformations. This process, in a chain reaction, affects the environmental and socio-economic conditions of the region. Though the Government of India has enforced the Environment Protection Act (MOEF,1986), declaring coastal stretches as Coastal Regulations Zone, due to lack of coordination among various agencies, implementation of the law has been ineffective. In addition, remedial measures (often not properly engineered) lead to complications which have a direct bearing on the socio-economic conditions of the coastal population. This article highlights various natural coastal activities initiated by processes which are either man-made or natural on the environmental and socio-economic scenario of the coastal states of Tamilnadu and Kerala.

COASTAL CHARACTERISTICS

The Tamilnadu Coast

Tamilnadu has about 1000 km long coast with an estimated population varying from 6000 in some coastal towns to four million in mega districts. Chennai (earlier Madras), the capital of Tamilnadu has an estimated population of about 6.5 million. Along north Chennai coast, extending from the fisheries harbour to Pulicat Lake (Figure 15.1), a population varying between 14,000 and 30,000 live in the coastal region. In the past few decades, Tamilnadu coast has witnessed an increase in industrial growth for exploration and exploitation of its marine resources, resulting in the migration of inland population toward coastal towns. Many new structures have been constructed to supplement the industrial requirements.

The coastal stretch of Tamilnadu exhibits varied geographical features interlaced with river systems, and coastal eco-systems like wetland, coral reefs, mangroves etc. In addition, the geographical configuration of the island state of Srilanka provides a well protected water body viz. "Gulf of Mannar" and "Palk Bay," leading to a conducive atmosphere for growth of marine organism (MOEF, 1992). Abundant supply of fertile sediment by the river systems enriches the coastal belt resulting in expansion of agricultural land along the river banks and delta region.

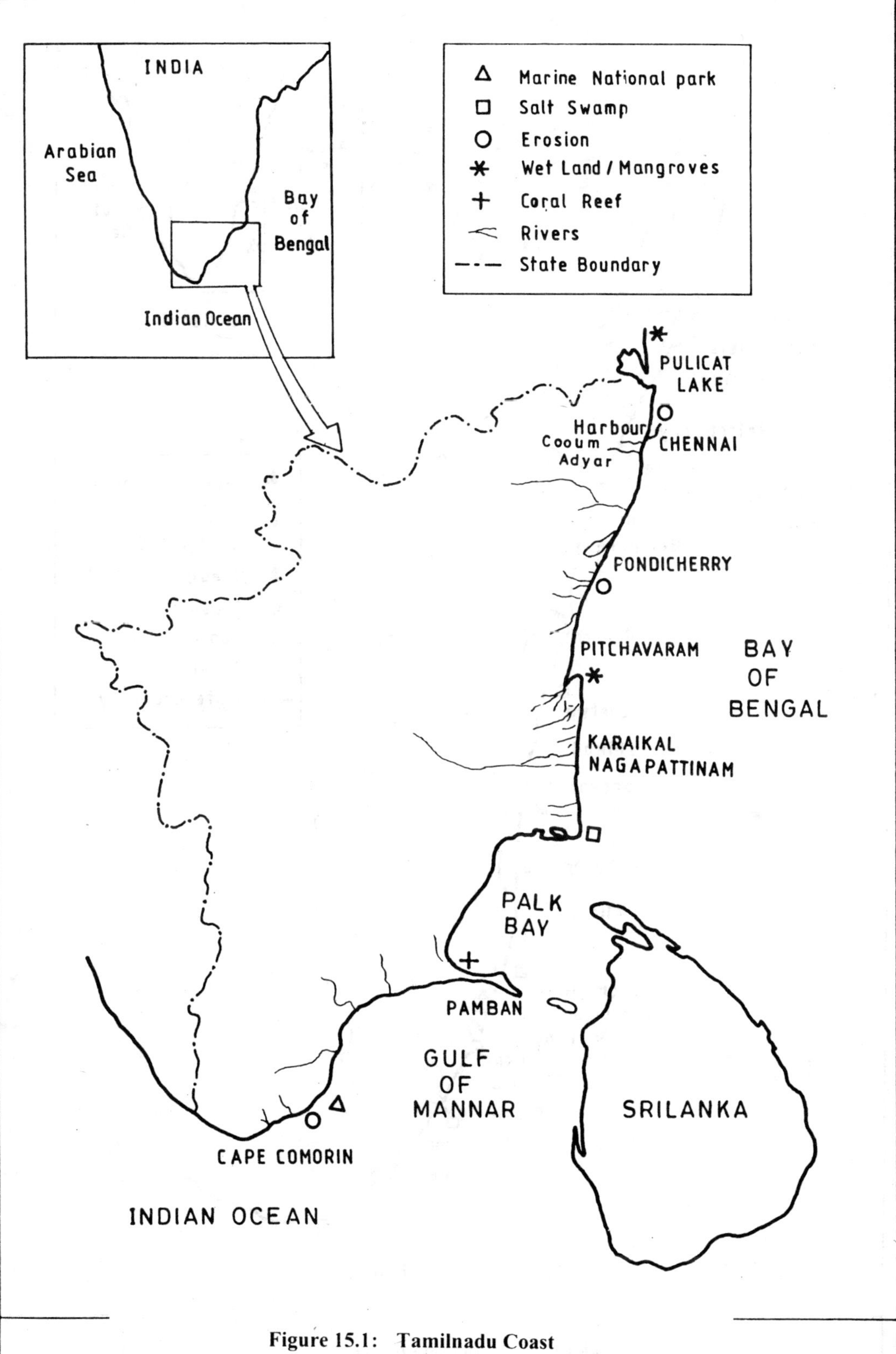

Figure 15.1: Tamilnadu Coast

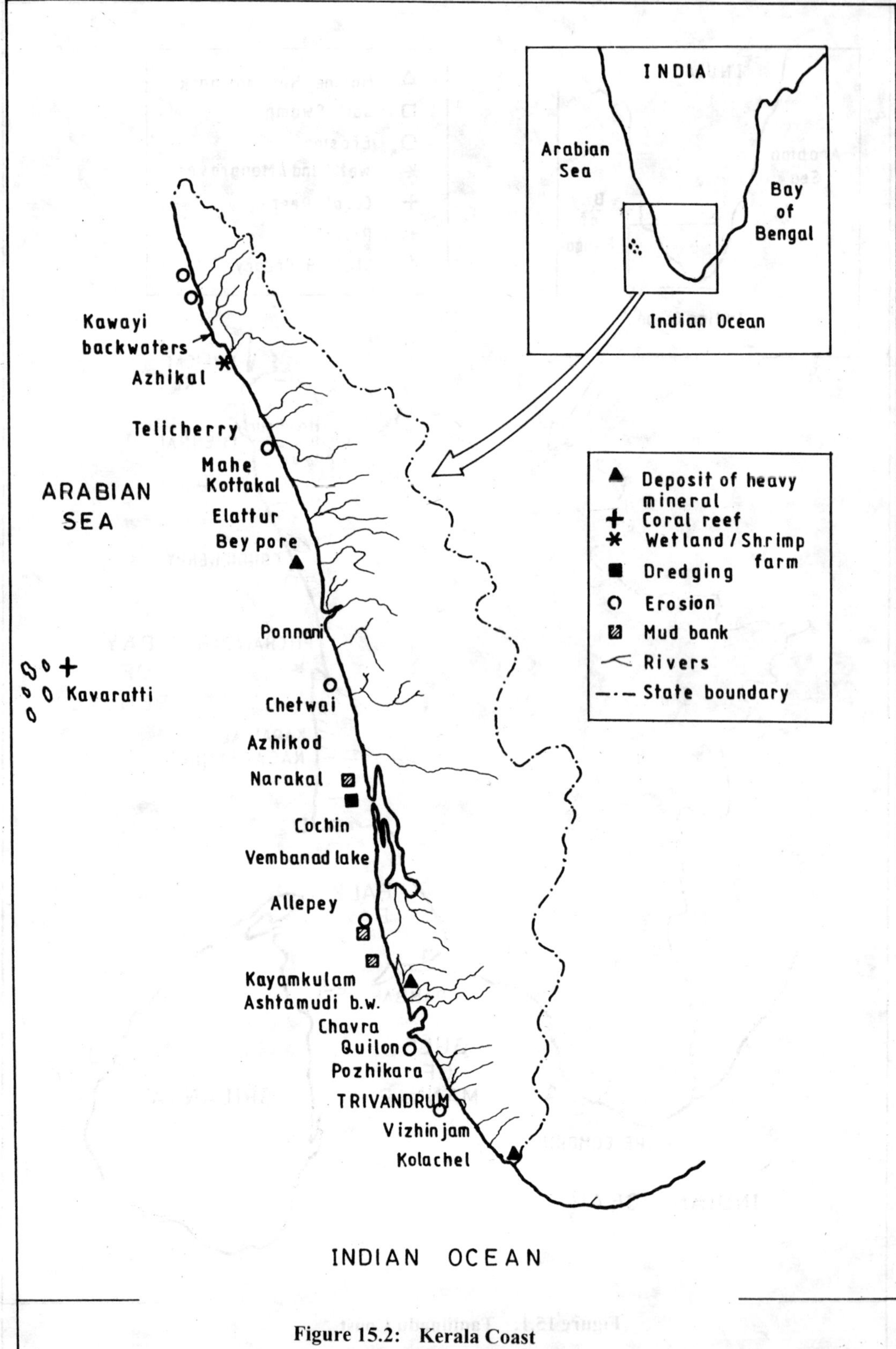

Figure 15.2: Kerala Coast

The geographical position of Tamilnadu coast makes the region vulnerable for wave attack both during northeast and southwest monsoon seasons. The region can receive waves as high as 8 metres during the northeast monsoon and during the southwest monsoon maximum wave height varies between 4 and 5 metres. A very important phenomenon associated with the coastal region of Tamilnadu is that a large quantum of sediment (about 0.6 million cubic metres per year) is transported along the shore by waves especially during southwest monsoon causing considerable modifications to the coastal characteristics. Though the northeast monsoon lasts for a few months in a year compared to the southwest monsoon, the coastal deformations produced due to the former are more damaging. For example a coastal stretch of about 10 kms north of fisheries harbour at Chennai is repeatedly threatened during northeast monsoon and the process continues. With the result, the existence of vital road link (East-coast highway) is challenged by the fury of the waves.

The East-coast of India is characterised by a phenomenon known as storm surge wherein water mass is literally pushed towards the coastline and held in its position by the wind system. This process results in inundation of coastal region and when associated with the tide and wave conditions can play havoc with the coastal establishments. A couple of coastal pockets in Tamilnadu are prone to such effects.

The Kerala Coast

The state of Kerala has about 570 km long coastal belt with population density varying between 750/km^2 inland and above 6000/km^2 along some coastal districts. Fishing industry holds the economy of Kerala state as it contributes to about 25% of total fish production in India. The fishermen population is about 0.9 million spreading over 200 villages. Kerala coast displays a wide range of geomorphological and environmental characteristics that is unique in nature. The coastline is intercepted by as many as 30 minor and intermediate rivers which discharge either directly into the sea or are connected to coastal water bodies technically known as "backwaters". Figure 15.2 shows the coastal region of interest. Unlike the East Coast, this part of coastal region experiences southwest monsoon on a large scale and occasionally the effect of northeast monsoon is felt. It has been observed that the coastal region experiences waves with a significant height ranging between 3 and 3.5m during June to September which has devastating effect along the coastline. Due to the severity of the wave climate many coastal stretches are subjected to erosion. Recorded observations indicate that the maximum wave height can reach upto 6.5m during monsoon season and upto 2.5m during non-monsoon period.

In contrast to Tamilnadu coast, numerous coastal features including bay formations, backwaters, steep cliffs etc. make Kerala coast more popular among scientists especially the coastal engineers. It is reported that this region experiences very low sediment transport (of the order of 0.07 million cubic metres) except for the southern part of Kerala where the transport rate is high (Felix Jose et al., 1997). Speciality of this coast is the deposits of heavy minerals and occurrence of a natural phenomenon known as "mud banks".

COASTAL PROCESSES

Coastal regions display a variety of processes inducing either a detrimental or beneficial effect on the coastal inhabitants. It is said that the coastal processes are always natural and the activity which initiates the processes can either be man-made or natural. Though the natural activity prevailed for billions of years, a perfect understanding between the processes and the coastal region left the coastal zone unspoilt. Ever since the mankind had started encroaching on the coastal belt, the problem of coastal imbalance emerged leading to complications. A list of natural and man-made activities which affect the coastal region are given below :

Activity	Processes
(i) Natural activities and associated processes	
(a) Waves-coast interaction	Coastal deposits and erosion
(b) Storm	Storm surge
(c) Tides	Sea level oscillation
(d) Wave-Seabed interaction	Flocculation of clay particles and formation of mud bank
(ii) Man-made activities and associated processes	
(a) Construction of coastal structures	Erosion and deposition
(b) Over exploitation	Mangrove degradation
(c) Human interference with backwater system	Contamination of water bodies
(d) Mining	Sand depletion
(e) Exploitation of living resources	Reduction in living organisms

The processes detailed above, and their environmental and socio-economic impacts are described in subsequent paragraphs.

Natural Processes and their Impact

Coastal deposits and erosion due to waves

The coasts of Tamilnadu and Kerala possess many river systems which act as a main source of sediment supply to the coastal region. These sediments, upon reaching the coasts are set in motion along the shore by the combined action of waves and river discharge leading to coastal deposits such as (i) Beach cusp and crescent shaped submerged bar (ii) Sand spit (iii) Longshore bar (iv) Cuspate foreland and (v) Tombolo.

These formations are shown in Figure 15.3 and as the emergence of sand spits is a common feature along the coast, their influence on environmental and socio-economic aspects are briefly given below.

Surfacing of sand spits at the end of monsoon season is a customary affair for the coastal rivers in Tamilnadu. These sand spits deflect the path of the river body at the confluence region resulting in (i) partial blocking of river mouth associated with formation of shoals at the entrance thus decreasing the water depth (ii) deformation of the adjacent coastal boundary.

Some socio-economic impacts of the processes are as follows:

(a) Increase in salinity level in the pond waters of the adjacent land-based agricultural and aquaculture farms causes substantial reduction in annual produce.

(b) Lack of oxygen in the water body and steep increase in salinity reduces riverine living organisms.

(c) Navigability of the waterway is restricted due to formation of shoals at the entrance of the river forcing fishermen to take the risk of grounding their vessels in these shoals.

(d) Migration of river mouth implies loss of beachfront and at times the inhabitants are compelled to find temporary shelter elsewhere till the natural beach re-building process is completed. Obviously this makeshift arrangement creates unquantifiable suffering to them.

(e) Stagnant water in the river system creates a breeding ground for mosquitoes causing health problem to citizens residing along the river banks (Adyar and Cooum rivers in Chennai, for example).

In case of Kerala coast, deposits of heavy minerals by streams reduce the water depth in the shorefront. As these mineral deposits cannot easily be transported by waves, the flexible sandy beaches on the neighbourhood take the brunt, often leading to loss of beachfront.

Storm surge

The East Coast of India is prone to frequent cyclonic storms which originate in the Bay of Bengal as low pressure zones and move towards the Indian Coast. At times these low pressure zones known as "depressions" intensify and grow into a full fledged cyclonic system accompanied by strong wind. The effect of a cyclone approaching the coast is to cause an abnormal rise in the sea water level adjacent to the coast and this phenomenon is termed as "storm surge". Figure 15.4 shows tracks of storms and depressions in the Bay of Bengal which were observed in the month of November between the years 1961 and 1970.

Extracts from the Atlas of India, Meteorological Department (IMD, 1979) indicate that there were about 1500 cyclonic disturbances during the period between 1877 and 1970, out of which 514 were either storms or severe storms. Of the 514 storms and severe storms observed in Bay of Bengal about 51 storms crossed the Tamilnadu coast. Table 15.1 indicates the storm surges that had occurred in the past (between 1955-1993) suggesting that the coastal stretch near Nagapattinam is most vulnerable for the phenomenon. The surge heights observed varied between 3 and 6m with an inland penetration upto 8 km.

The effect of storm surge is to temporarily rise the water level and if this extreme phenomenon coincides with the occurrence of high water level during a spring tide, the combined action produces an alarming situation such as

(i) inundation of low lying land area adjacent to the coastline.

(ii) free access for massive waves (with a ht. exceeding 2 m) to reach the coast.

Unless a pre-evacuation warning is given to the inhabitants in coastal areas, it will lead to sizable destruction of coastal resources and population. As the coastline along

Karaikal and Nagapatnam in Tamilnadu normally experiences the cyclonic conditions producing a storm surge crisis management programmes by the Government of Tamilnadu were initiated in the recent years (Mani, 1997). As far as Kerala coast is concerned, though it had experienced a few cyclonic depressions they were not strong enough to generate a storm surge.

Table 15.1: Storm Surges along Tamilnadu Coast

Year and Regions Affected	Max.Wind during Cyclone (kmph)	Surge Height	Inland penetration	Damages, loss of human life
Nov. 1952 South of Nagapatnam		3m	5 miles	Several thousand died
Dec. 1955 Rajamadam (Tanjore Dt.)	200	10-15'	2 to 5 miles	500 human loss
Oct. 1963 Cuddalore	139	20'		
Dec. 1964 Rameswaram Island	193	3 to 5m		1000 people died
Dec. 1967 Nagapatnam	130		Tanjavur district between 10° N & 11.3^{o}N affected	
Nov. 1978 Ramanathapuram	212	3 to 5m		10 deaths in India
Nov. 1991 Near Karaikal	89		200 to 250m	
Dec. 1993 Near Karaikal	133	3 to 4m	2 km	

Sea level oscillations

Sea level rise has been the focal point for many scientists in the last few decades, with an outcome indicating that the waterlevel would rise (globally) by about one millimetre upto a few centimetres in the next half century. Though this long-term rise in water level is of concern, priority goes to the daily fluctuations in the water level that the coast experiences due to gravitational attraction between the celestial bodies. The phenomenon has pronounced effect on the coast of Kerala as it possesses large number of inlets and backwaters. As the useful land area along this coastal region is restricted, increase in water level further reduces the effective land use. In addition, continuous tidal exchange combined with seasonal discharge of fresh water weaken the soil strata which indirectly imposes restrictions on the construction of residential and eco-friendly industrial complexes. Supply of sediments from the river systems to the backwaters and influx of sea water has reduced the cultivable land area thus adding difficulties to the coastal community depending largely on agricultural produce. In some locations sea water inundation has compelled many of the inhabitants to switch over to other professions from traditionally practiced agricultural activity.

Mud banks

A very interesting phenomenon that occurs along the coast of Kerala is the formation of mud banks, known to emerge only during southwest monsoon (Sassikumar, 1997). Action of waves on a clayey bed surface wash out the top layer and the large waves keep them in suspension to form mud banks. This coat of mud, kept in suspension near the water surface, at a distance offshore acts as a natural barrier and protects shoreline in the vicinity from large waves. In the process, the unprotected region of the coastline takes the brunt of the waves leading to (i) coastal deformations and (ii) coastal inhabitants forced to find alternate sites for settlement.

Processes induced by Man-made Activities and their Impact

Coastal erosion due to construction of structures

Many examples can be cited along the coasts of Tamilnadu and Kerala, wherein coastal structures like harbours and protective rubble walls have been constructed leading to coastal inequilibrium. Finest example is the influence of Madras harbour on the coastline. Stage-wise construction of the Madras harbor obstructed the natural sediment transport and resulted in the formation of the famous Marina beach whereas the shoreline north of the Madras harbor suffered pronounced erosion (Figure 15.5).

It has been estimated that about 500m of land has been lost in a period of about 100 years due to the construction of Madras harbour. As the shoreline was continuously receding, attempts were made by the State Government to protect a coastal stretch of about 3 kms and further north the coastline was left unprotected. Due to construction of rubble wall, increase in water depth in the nearshore area resulted in large waves reaching

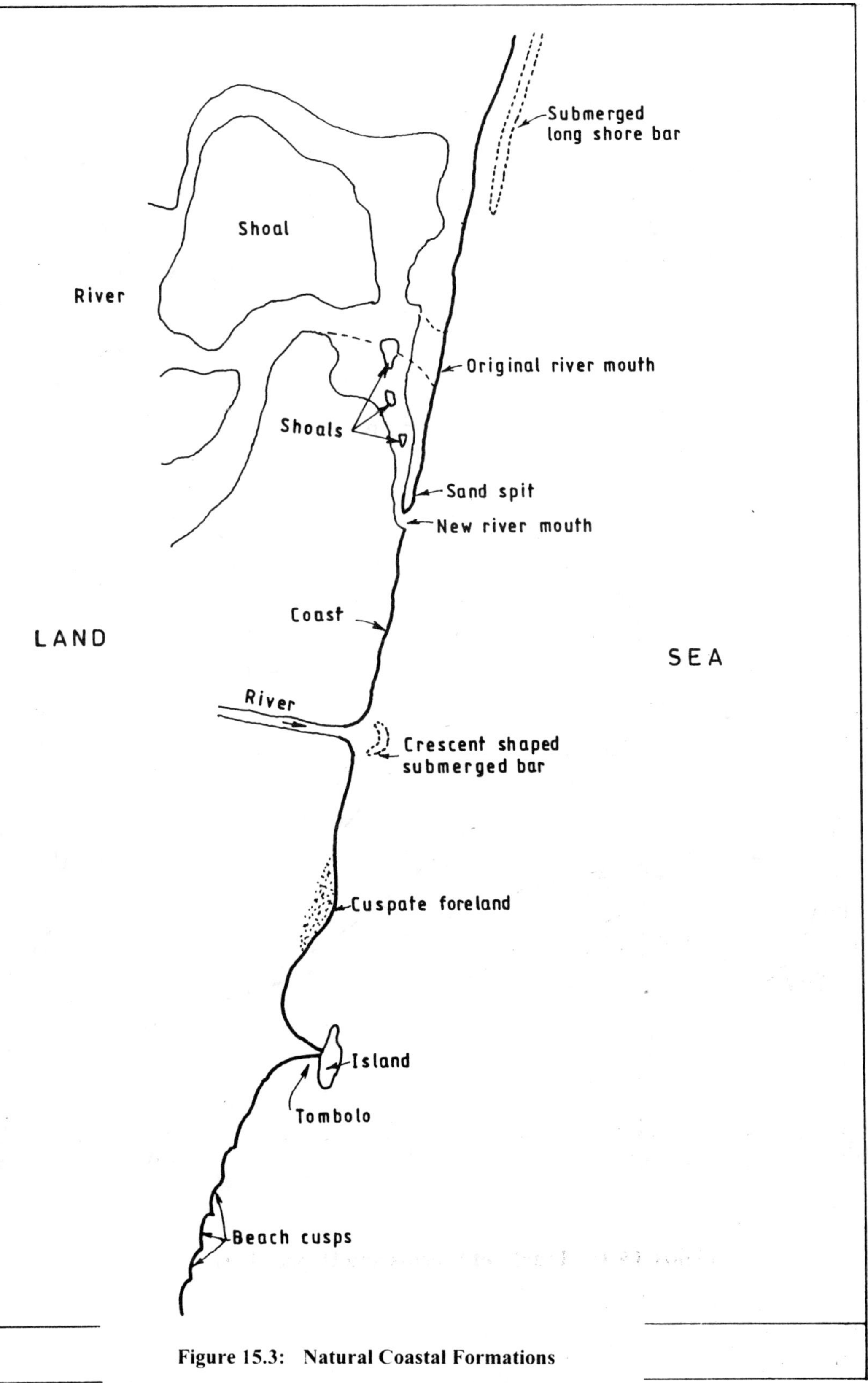

Figure 15.3: Natural Coastal Formations

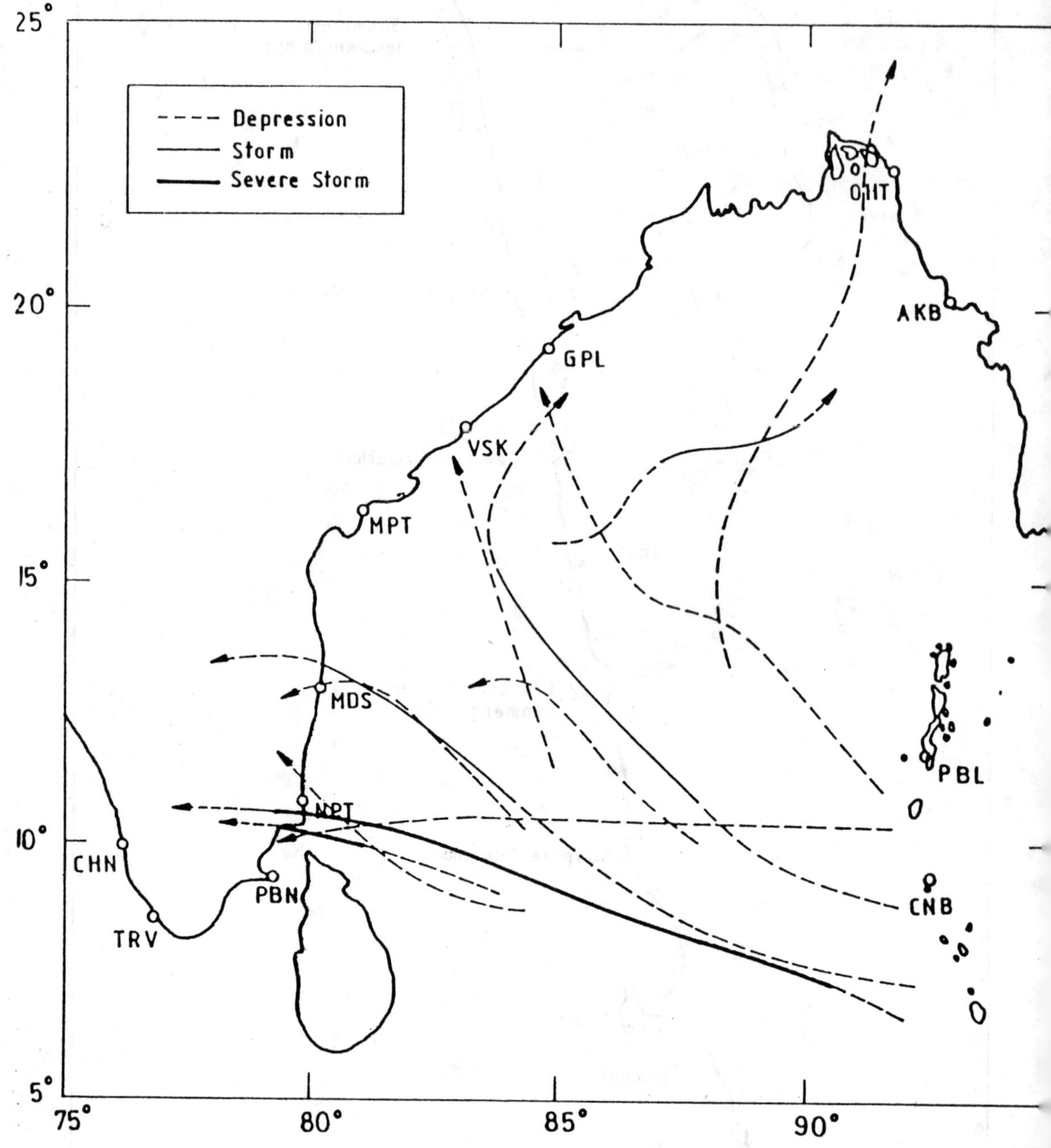

Figure 15.4: Tracks of Storms and Depressions

the coastline which in turn affected the hutments of coastal fishing community as the spray from waves was enormous. In the recent years, unprotected region of the shoreline has been experiencing severe erosion resulting in disruption to vehicular movements along a national highway running adjacent to the coastline. Temporary measures to check the erosion were not successful as they were not designed and executed based on engineering solutions. The situation as on today is that the fishermen community along the coastal stretch extending for about 3 km have been evacuated and rehabilitated elsewhere. Similarly a couple of small scale industries faced the brunt of the waves compelling the management to close down. The process of erosion is still continuing and threatening the vital coastal highway and adjoining large size industries. Unless the erosion is checked with proper engineering solution, loss to the property may run into millions of rupees apart from generating following additional problems.

(i) Loss of land, thereby increasing the pressure on densely populated metropolitan city of Chennai.

(ii) Depletion in potable water due to sea water ingression.

(iii) Permanent loss of settlements for fishing community due to denudation of beach front.

Recently, construction activities have commenced at about 15km north of Madras harbour for development of a satellite port which certainly would have some influence on the existing pattern of sediment transport and on the coastal equilibrium. An unprecedented activity emerging along this part of the coastal belt is the mushrooming of container depots. Frequent movement of container carriers tend to baffle both the fragile coastal highway as well as the coastal community. Vast stretch of coastal belt along Kerala has been protected to prevent coastal erosion leading to unforeseen complications in the living habits of the local population. Though these construction activities are beneficial in short term, their inherent characteristics produce ill-effects to the coastal population from the point of view of the following.

(i) Acceleration of coastal erosion process on the downdrift side of the structure,

(ii) Increase in water depth enabling higher waves to reach the coastline threatening the population along the coast,

(iii) Imposing restrictions on the free access by fishermen to the sea,

(iv) Acute shortage of drinking water due to seawater ingression,

(v) Effect on plantations and other crops due to non-availability of fresh water thus, reducing the yield,

(vi) Forcing the fishermen to migrate in search of good beach front for landing their vessels and for venturing into the sea and

(vii) Imposing restrictions on coastal transportation.

Mangrove degradation due to over exploitation

Mangrove spread in Tamilnadu, popularly known "Pitchavaram," covers a total area of about 1,100 ha, consisting of 51 islets with their size ranging from 2 to 10 sq.km. The area has a number of channels, creeks, and rivulets. About 50% of the total area is covered with mangrove forest, 40% with waterway, and the remaining 10% with sand and mud flats. Out of 30 species of plants, 20% of the species are woody trees, and the waterway serves as a potential spawning area and nursery grounds for a variety of finfish and shellfish. It is estimated that the mangrove leaf litter production amounts to about 7.5 tonne/ha/year.

Some of the Impact of Degradation of Mangroves are as follows:

(i) In the coastal area, especially near mangrove forest, prawn culture farms established in large scale produce effluents (with organic and inorganic pollutants). These effluents damage the ecology of the system.

(ii) Collection of postlarvae in the mangrove area by the fishermen, for distribution to the aquaculture farms, has caused depletion of natural shrimp production, and hence, the unit stock.

(iii) Cutting of trees for fuel and overgrazing by cattle has led to soil erosion and stunted the growth of plants. (It has been estimated that the population living within the 20-km belt of the coastline requires nearly 2 million tonnes of fuel wood annually).

(vi) Dumping of wastes by tourists who throng the area for recreational purposes, has caused pollution.

(v) Out of the total mangrove spread, only 50% of the area is in good condition and important species are in danger of extinction.

(iv) Decrease in nutrient level in the region which otherwise would have a stock of nutrient rich sediment supplied from the low lying land.

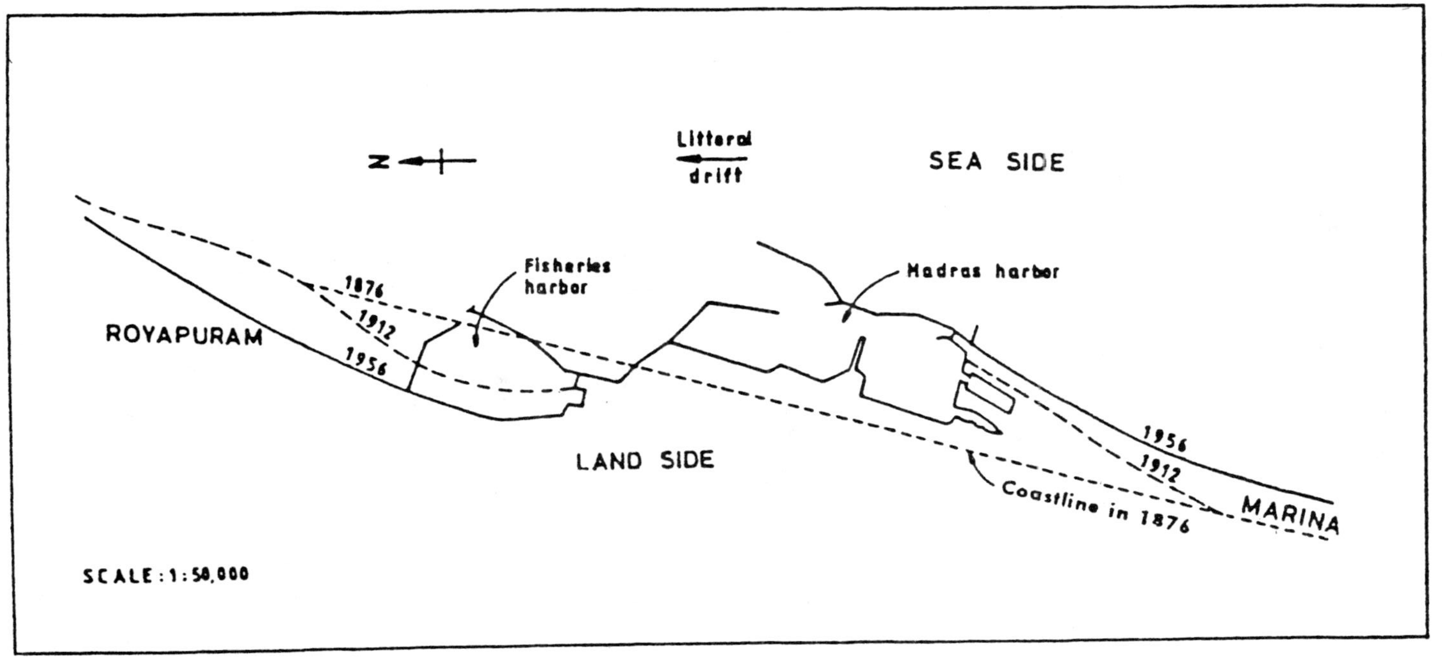

Figure 15.5: Deposition and Erosion Pattern near Madras Harbour

Human interference with backwater system

It has been reported that ecologically sensitive backwater systems of Kerala, and the fish-life are under threat from various development schemes, such as construction of buildings resulting in increase in land price and reclamation of the backwaters, dredging of waterways in the backwater for navigation purposes, conversion of wetlands into aquaculture farms etc. The main target has been the Vembanad Lake near Cochin, the largest brackish water body which has shrunk to a quarter of its original size. It has been alleged that the State Government decision to introduce oil palm cultivation in the Kuttanad rice belt would threaten the wet lands. The development of Cochin International Airport, sports and tourist complexes, development of Cochin Port and minor dams across Periyar river are expected to have an adverse effect on the environment of the backwaters.

At Cochin Port, to maintain the approach channels with a water depth of about 10 to 13 m, dredging is regularly carried out (Rasheed and Balchand, 1997). This operation, eventually influences the turbidity causing concern to environmentalists as it reduces productivity level of marine life. In addition the flora and fauna of the area face threat which disturbs the ecosystem.

Extensive area along the north Kerala coast covering an area of about 540 sq.km reveal that in the past three decades sizeable reduction in wetlands has occured (Nandakumar and Salim, 1997). In general it has been observed that mud flats and mangrove regions converted for aquaculture in the past have now been left unutilised due to outbreak of viral diseases. The socio-economic impact due to development of aquaculture farms lead to

i. conversion of fertile agricultural land into aqua culture farms,
ii. displacement of manpower from various places for manning these farms,
iii. use of chemicals and nutrients contaminated the fertile land as well as the ground water quality,
iv. outbreak of virus related diseases lead to collapse of the system and it was not possible to revert back to agricultural farming and
v. the process of generated unemployment.

Sand depletion due to mining

Kerala coast possesses a couple of pockets rich in deposits of heavy minerals varying from magnetite with a density of about 5.1g/cm^3 to Hornblende with a density of 3.2g/cm^3 (Prakash et al., 1997). It is fascinating to note that these deposits are associated with medium sized streams with their runoff occuring through steep gorges of the adjacent hilly terrain. Due to the heaviness of the material, they tend to get deposited at the stream mouth and transported at a slow rate along the shore even when strong waves reach the coast. Continuous extraction of heavy mineral warrant for mining of these deposits in large quantities, leading to morphological changes in nearshore profiles which in turn pave way for large waves to have access to thickly populated beach fronts.

Reduction in living organisms due to over exploitation

Over exploitation of living resources like fishes and corals lead to an imbalance in the marine life and decrease in the marine population at times destroy the species which contribute to unsustainable growth. Exploitation of corals leave the coastal zone exposed to the severe wave climate which directly influences the coastal population. The coral and coral reef organism are highly susceptible to environmental stress, mainly pollution. Increased industrialisation in coastal areas, off-shore mining and oil spills increase the risk of large scale pollution. In addition, mining, dredging, harbour and off-shore constructions, shell and coral collections, establishment of recreational facilities along the beaches are the important factors that affect the sustainability of reefs. The most serious threat to Coral reefs is the excessive and indiscriminate exploitation which caused irreversible damage, as seen in the Gulf of Mannar and at Tuticorin. Removal of corals and dredging of the lagoon floor had led to erosion of shorelines in Kavaratti island off Kerala coast.

Disturbance to the coral reef has the following impact on coastal environment (ESCAP, 1995):

(i) decrease in the marine fishes as they are mainly dependent on coral reefs for support,

(ii) the beaches are left unprotected against the storm surges and waves, and

(iii) reduces the possibility of developing pearl culture and other associated industries.

CONCLUDING REMARKS

Delicate coastal region is constantly threatened, ever since human interference exceeded the tolerance limit. Detrimental effects arising out of the situation has been manifold, which would certainly restrict the privileges that our future generations should enjoy. This has been the concept of sustainable development and it is the duty of present day citizen for keeping the natural resources intact so that the mankind can reap the benefits for years to come.

ACKNOWLEDGEMENT: The author would like to express his sincere thanks to the authorities of Indian Institute of Technology, Madras, Chennai for their support. The author also expresses his thanks to Ms.K.Lakshmi for the support extended by her in typing this manuscript.

REFERENCES

Felix Jose, Kurian N.P., and Prakash, T.N., (1997), "Longshore Sediment Transport along the Southwest Coast of India", Second Indian National Conference on Harbour and Ocean Engineering, Thiruvananthapuram, pp.1047-1053.

Mani, J.S., (1997), "Report : Coastal Zone Management Plans and Policies in India", Coastal Management, An International Jr. of Marine Environment, Resources, Law and Society, Vol.25, No.1, Jan-Mar, pp.93-108.

Nandakumar, D., and Salim, M.B., (1997), "Coastal Resource Management: The Case of Shrimp Culture Development in Kannur District", Second Indian National Conference on Harbour and Ocean Engineering Thiruvananthapuram, pp.1283-1290.

Prakash, T.N., Kurian N.P., and Felix Jose, (1997), "Heavy Sand Concentration in Relation to Hydrodynamic Processes along Manavalakurichi, SW Coast of India", Second Indian National Conference on Harbour and Ocean Engineering, Thiruvananthapuram, pp.1099-1107.

Rasheed, K., and Balchand, A.N., (1997), "Dredging Impact Assessment (DIA) at Cochin Port", Second Indian National Conference on Harbour and Ocean Engineering, Thiruvanthapuram, pp.586-594.

Sassi Kumar, P.K., (1997), "The Mud Banks of South-West Coast of India and its Effects on the Coastline", Second Indian National Conference on Harbour and Ocean Engineering, Thiruvananthapuram, pp.1191-1204.

IMD, (1979), Indian Meteorological Department, "Tracks of storms and depressions in the Bay of Bengal and the Arabian Sea", Puna, India : Indian Meteorological Department Publications.

MOEF, (1986), "Ministry of Environment, Forests and Wildlife", Government of India, The Environment (Protection) Act.

WWF, (1992), "India's Wetlands Mangroves and Coral Reefs", published by World Wide Fund for Nature India, Ministry of Environment & Forests, New Delhi.

ESCAP, (1995), "Planning Guidelines on Coastal Environmental Management", Economic and Social Commission for Asia and the Pacific, United Nations, New York.

16

NATURAL RESOURCES MANAGEMENT OF ANDHRA PRADESH COAST THROUGH REMOTE SENSING TECHNIQUES

Gautam N. C., Jayanthi S. C., Ravi Shankar G. and Raghavswamy V.

ABSTRACT

Coastal environment is vital for the prosperity of any nation. It often consists of - highly productive agricultural lands, densely populated cities, heavily industrialised zones as well as ports and harbors which act as nerve centres of a nation's economy. Its' protection is the prime responsibility of the planners, decision-makers as well as the nation's intelligentsia. Because of its preferred status, it is often beset with many environmental problems resulted by manmade activities. The ecological degradation can be perceived through forbidding levels of atmospheric pollution, polluted sea waters, silted up shorelines, intrusion of hard water into interior land portions, etc. To minimize the environmental damage and also to conserve the area's present productivity levels for future generations, its planning is imperative. Some solutions in this direction are offered through satellite remote sensing technology for the conservation and betterment of land and water part of this biosphere.

INTRODUCTION

India has a lengthy coastline extending from Gulf of Kutchch in the west to Sunderbans in the east touching the country's borders with Bangladesh. The eastern coast of India, popularly known as Coromandal coast occupies a prominent position in the history as well as economy of the country. The Coromandal coast is basically possessed by four States viz., West Bengal, Orissa, Andhra Pradesh and Tamil Nadu. As India slopes from west to east, all the major rivers in the country too flow in the same direction, draining into the Bay of Bengal. Thus, some of the major river deltas of India are embedded in the Coromandal coast. These river deltas with their fertile soils often have shaped the economies of these States for good by boosting agriculture production. Besides this conventional bounty, the coastal plains are also profited by other means like growth in aquaculture sector, oil exploration, accelerated industrial growth, etc. But, all the same, there exist some detrimental effects of this rapid growth in various sectors. Some of the major spin-offs are high population densities, need for improved communication and transportation networks, threat of environmental pollution etc. Thus, it becomes imperative for both planners and decision-makers to study the developmental scenario in a holistic sense and the corresponding threat to environment before embarking on implementation works.

For the sustenance of future generations of mankind on this globe, frugal and judicious usage of resources is a must. Thus, resources planning attain high importance in the overall planning scenario; commanding an equal place to that of economic or human resource planning. Resource planing calls for structured and up to date database without any bias. Usually the success or lack of it of a given plan depends on the quality of the input database. Conventional methods of resource information collection, mapping are becoming prohibitively costly, time consuming and are error prone. Of late, the technology of satellite remote sensing is fast emerging as a viable and cost effective alternative to conventional technique. It is also proving to be useful furnishing the information in a timely manner which the earlier technologies couldn't match. Equipped with these advantages, remote sensing technology is fast making forays in the resource management field.

Evolution of Indian Remote Sensing Program

The late 1960's witnessed the initiation of remote sensing program in India, which are mostly confined to the studies using aerial sensors. This was followed by systematic and fabulous efforts in aerial flights, development of remote sensors, data acquisition from Landsat series and SPOT satellites, planning and implementation of experimental earth observation satellite missions and setting up of ground based data reception, processing and interpretation facilities. The launch of Bhaskara I and II in the early 1980's gave a boost to the user community. This was followed by the conceptualization of Indian Remote Sensing Satellites (IRS) series for the operational application of remote sensing in natural resource management. The indigenously built IRS-1A and IRS-1B satellites has opened the floodgates on the flow of resource information for use on operational basis by the user community.

Considering the requirements for different applications with higher spatial resolution, the second generation satellites with frequent revisit capabilities, stereo vision and onboard video data recording facilities, the IRS 1C and 1D satellites with identical payloads were launched which are operational from 1996 and 1997 respectively. Also, the Department of Space, has set up the mandate for the launch of IRS P4 (OCEANSAT – 1) with payload specifically tailored for the measurement of physical, biological and oceanographic parameters. It is also planning launch of the IRS P5 (CARTOSAT) specifically for use in cartographic and terrain modeling applications, IRS P6 for agriculture applications, IRS 2 series (OCEANSAT 2/CLIMATSAT 1/ATMOS 2), an integrated mission to cater to global observation of climate, ocean and atmosphere.

Operationalisation of Remote Sensing Applications

The Government of India in the year 1985 has approved the establishment of the National Natural Resources Management System (NNRMS) to address the issues relating to natural resources management in the country. Agriculture, bio-resources and environment, geology and mineral resources, ocean resources, water resources and training in remote sensing technology are the thrust areas identified under this programme. Five Regional Remote Sensing Service Centres (RRSSC) have been established under NNRMS to cater to the needs of the five regions in the country. Also, establishment of State Remote Sensing Centres for each of the state was encouraged by the Dept., of Space.

Large scale operational projects and to some extent a few projects on experimental basis have been taken up by various agencies using remote sensing data. An attempt has been made to chronicle here a few examples depicted the utility of remote sensing technology for the improvement of resources position in the coastal landscape of Andhra Pradesh.

Characteristics of Coastal Districts

The districts of Srikakulam, Vizianagaram, Visakhapatnam, East Godavari, West Godavari, Krishna, Guntur, Prakasam and Nellore constitute the coastal districts of Andhra Pradesh (Figure 16.1).

Geologically, the East Coastal plains predominantly consist of Recent and Tertiary alluvium. Patches of Archaean gneisses and sandstones etc. are also found along the coast. In Andhra Pradesh, Tertiary formations are found in East and West Godavari districts and in small areas of Nuzvid tehsil of Krishna district. Parts of Rajahmundry and Peddapuram tehsils in East Godavari, Eluru and Tadepalligudem tehsils in West Godavari tehsils contain Tertiary formations of clays, useful for ceramics developed near Rajahmundry. Some patches of laterites are also found in Visakhapatnam in the south, more in the mainland in Guntur, Krishna, East and West Godavari districts.

The coastal plain comprises of rich fertile soils in the deltas of Godavari, Krishna and Pennar rivers. Godavari is the largest perennial river in Peninsular India. After crossing the Eastern Ghats, it emerges at Polavaram into the coastal plain. The width of the river is over 3 km, at Rajahmundry and about 6 km at Dowlaiswaram. Below Rajahmundry it splits into the Gautami, Vashista and Vainateya branch which form its delta. The three branches join the sea at Yanam, Narsapur and Razole respectively.

The Krishna is the second important river. It is superimposed across northern end of the Cuddapah ranges where the gradient is 0.7 m per km. Near the sea, the gradient is 0.15m per km. It flows into two branches near Paugedda in Krishna district enclosing the island of Diwi, and 16 km downstream splits into three branches. The Vamsadhara and Nagavali are other notable streams flowing in Srikakulam district. And the Pennar in Nellore district.

The region is abounding in alluvium. Red soils, black soils and laterites are also found as transported soils. Alluvial soils are mostly found in river valleys, deltaic tracts and along the coastal area; their composition and textures vary with the geological nature of the catchment area. These are again of two types – coastal alluvium and deltaic alluvium. These are exceptionally fertile and are highly valuable for agriculture, especially paddy. Red soils occupy considerable parts of Srikakulam, Visakhapatnam

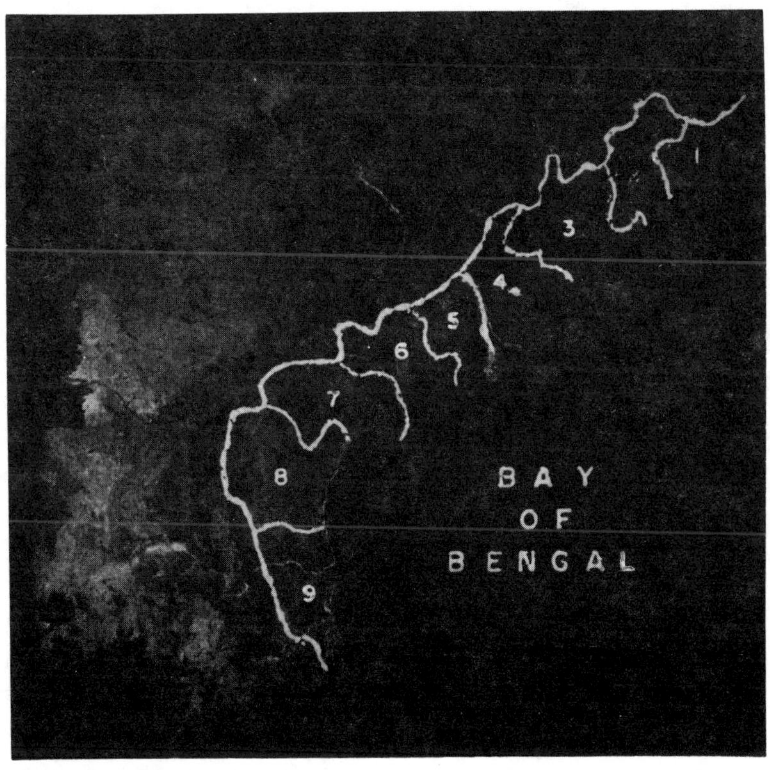

1.	Srikakulam	6.	Krishna
2.	Vizianagaram	7.	Guntur
3.	Visakhapatnam	8	Prakasam
4.	East Godavari	9.	Nellore
5.	West Godavari		

Figure 16.1: Satellite Image Mosaic of Andhra Pradesh with the Coastal District Boundaries

and East Godavari districts; and also in small areas of Krishna, Guntur and Nellore districts. Laterites are tropical soils formed by the decomposition of gneiss. The most important components for lateritic formations are iron, alumina and silicic acid as primary material of the parent rock, seen distributed in Nellore, East and West Godavari districts. Black soils, rich in lime, magnesium and aluminium are observed in parts of West Godavari, Krishna, Guntur, Prakasam and Krishna districts adjoining coastal alluvium.

Status of Remote Sensing Applications In Coastal Zone

One of the earliest works that was executed for the State of Andhra Pradesh using satellite remote sensing for mapping natural resources dates back to the year 1978 when all the natural resources were mapped on 1: 1 million scale using Landsat Multi Spectral Scanner (MSS) data. However, with the improvement in sensor technology over a period of time, each of the application discipline has evolved as a specialized subject. The areas in which significant headway was made are detailed below.

GEOLOGY

Remote sensing technology offers a wide scope of applications in the field of geological sciences like lithological discrimination and mapping, structural mapping, landform analysis, hydrogeology, mineral exploration etc. The synoptic coverage and multi spectral information provided by the remote sensing data have proved to be advantageous over conventional methods for geological mapping. The synoptic view help in visualizing the terrain as a whole and to understand the spatial relationship between different features. The multi spectral information obtained in the different wavelength regions of the electromagnetic spectrum enable differentiation and mapping of specific geological units.

The other aspect of this technology is in geomorphology. The landforms are excellently depicted on the satellite data. Stereo capabilities of the present day satellites like IRS 1C and 1D permit evaluation of slopes, relief and form. The dynamic changes in physical processes and resulting landforms (like river migration) are documented in remote sensing data and can be studied by virtue of temporal resolution of the data. In the advanced system, thermal images acquired during nighttime provide an excellent impression of regional landforms, because of the differences in nighttime temperatures, related to elevation, soil moisture, vegetation, etc. Space borne images gives a good perception on various landforms like beach ridges, palaeochannels, meander scars, natural levees, structural hills, denudational hills, inselbergs, pediments, sand dunes, etc. Analysis of landforms in combination with drainage enable lithological discrimination and structural mapping; and understanding the geological processes operating in the region.

Satellite remote sensing studies play a vital role in studying and monitoring geo-environmental conditions like zones suitable for aquaculture, salt industry etc. It also helps in understanding geomorphic processes such as shoreline changes, siltation of lagoons and estuaries, migration of river courses, degradation of mangrove forest, location of - ground water prospective zones, - trace minerals, - hydro carbons, zones suitable for aquaculture, salt industry etc. The study carried out around Kakinada in East Godavari district has revealed significant changes. Analysis of Landsat MSS data of 1973 has shown formation of three spits at the mouth of the Gautami River. Analysis of IRS data of 1989 indicates the removal of one of the spits by erosion and also the southern distributory of Gautami Godavari becoming less active (shrinking in width). A comparative study of the two data sets indicated change in the morphology of the spits, development of beach near Masanutippa and migration of the Gautami Godavari. The study of the topographical map published in the year 1938 revealed that southern distributory of Gautami Godavari was more active and a number of offshore bars were also present at its mouth. Similar studies in the Krishna delta revealed incipient growth of two spits on either sides of the Krishna delta, which have grown in size by 1988.

The other important and related field is identification of prospective zones of ground water occurrence. Factors controlling the ground water regime are manifold, which primarily encompass underlying rock types, geological structures, landform characteristics, recharge conditions etc. In conventional hydrological mapping, the emphasis is given to geology wherein individual rock type is taken as a base for evaluation of its aquifer characteristics vis-a-vis ground water prospects. Hence, for a better understanding of ground water prospects, it is meaningful to incorporate geomorphological, structural and hydrological information with geological information to derive the ground water prospective zones.

The spatial and temporal variation in rainfall, differences in the geology, relief/slope is the factor responsible for uneven distribution of ground water. This coupled with haphazard and unplanned exploitation resulted in shortage of this resource even for the basic needs and on the extremes resulted in water logging, salinity and flooding. In this context, finding groundwater sources for planning sustainable drinking water schemes assume great significance. Realising this, Government of India launched the National Drinking Water Technology Mission in 1987, with the main objective of providing safe drinking water to all habitats in the country. The Department of Space, Government of India was entrusted with the responsibility of preparing ground water prospective maps for all districts in the country on 1: 250,000 scale, which was successfully accomplished with active collaboration of Central/State Government organizations. As a part of this mission, hydro-geomorphological mapping for all districts along the East Coast was completed. These investigations constitute inventory of existing wells in the area for -ground water levels/fluctuations, aquifer material characteristics, water (quality) sample analysis, area of recharge (soil/rock type), location and quality of ground water discharge at surface, water availability for recharge through: precipitation, surface water-flows, tanks/reservoirs etc.

AGRICULTURE

Research efforts in the application of remotely sensed data in crop studies has resulted in the pre harvested acreage estimation of major crops like wheat, paddy, sorghum, soybean etc., and condition assessment of crops. Intensive efforts are on to develop remote sensing based yield forecasting models. Experiments are being conducted to assess crop water requirements also. As a part of a project work undertaken for the Ministry of Textiles, Government of India, crop acreage and condition assessment for cotton crop was taken up for Guntur and Prakasam districts, where there are black cotton soils that support growth of this crop.

The major emphasis in crop studies is to develop yield models based on spectral indices (vegetation index) of crop and agro-meteorological parameters. Because of lack of sufficient spectral database, the crop yields are being forecasted based on the analysis of historical yield data and assigning weightage for crop condition and other technological variables. This technique is being adopted for forecasting cotton and soybean crop yields since 1993 kharif season.

The space borne data requirement for crop studies has a serious limitation of cloud cover. The all-weather and all-time observation capabilities of microwave sensors offer great potential for crop identification and acreage estimation, especially during monsoon season. Utility of temporal Synthetic Aperture Radar (SAR) data for crop identification has been investigated in the East and West Godavari districts of Andhra

Pradesh. European Remote Sensing Satellite (ERS-1) SAR data acquired over the area on three dates during rabi season of 1992-93 processed using a 3 x 3 median filter and co-registered and a time composite is generated combining SAR data of December, 1992, January, 1993 and February, 1993. Various crops like early and late transplanted paddy, early and late planted tobacco, pulses such as green gram and black gram, plantation crops like mango and cashew, coconut and banana and mangroves were seen in different colors. Other features like water in river Godavari and built-up areas could be distinguished based on colors.

SOIL STUDIES

Information on soils, their capabilities and limitations is needed for various agricultural and non-agricultural purposes. Soil survey provides such information and also maps, which are essential to evaluate the soils, suitability to irrigation, specific crop, response to fertilizer application, for optimum land use planning etc. Traditional soil surveys are slow, subjective and laborious for preparing soil inventories. The systematic efforts in the application of satellite data in soil survey culminated in the development of methodologies to map soils on operational basis. The utility of remotely sensed data for soil resources mapping has been demonstrated by many studies (NRSA, 1994). In the preparation of soil map from remotely sensed data usually visual interpretation of False Color Composites (FCCs) method is adopted, as it is advantageous.

A study was taken up for the Tribal Welfare Department, Government of Andhra Pradesh on soil survey and agricultural land use planning encompassing the districts of Srikakulam, Visakhapatnam, Vizianagaram, East Godavari districts. The project aimed at preparation thematic maps on soils, hydrogeomorphology, slope, land use/land cover, land capability, problem soils maps and generation of agricultural land use plans on 1: 50,000 scale.

Space borne remote sensing data was found to be extremely useful in mapping and monitoring of various degraded lands like salt affected soils, eroded and jhumlands, waterlogged areas, ravinous areas, wetlands etc. Mapping of salinity and alkalinity was done for the south coastal districts of the State covering the districts of Krishna, Guntur, Prakasam and Nellore on 1: 50,000 scale for use by the Department of Agriculture for planning reclamation measures and improving agricultural productivity.

LAND USE/LAND COVER

The spatial information on land use/land cover and their pattern of changes is a necessary prerequisite for planning, utilization and management of land resources. Land use/land cover inventories are assuming increasing importance in various resources sectors like agricultural planning, settlement and cadastral surveys; environmental studies and operational planning based on agro-climatic zones. Information on land use/land cover permits a better understanding of the land utilization aspects on cropping pattern, fallow land, forest and grazing land, wasteland, surface water bodies etc., which is vital for developmental planning.

Much of the planning activities till recently were heavily relying on statistical information collected and compiled based on conventional sources, which of course has severe limitations. In the last two decades, satellite remote sensing techniques were employed for generating information at national level and on 1: 250,000 scale using Landsat MSS, TM data. However, new developments in the technology have considerably increased the potential use of satellite images for large-scale mapping.

Realising the importance of land use/land cover information, the Planning Commission, Government of India felt the need for having an latest information for the whole country on agriculture and other land use categories and their estimates for planning based on agro climatic zones. This national level task envisaged preparation of land use/land cover maps depicting a composite picture of kharif and rabi seasons at district level, using IRS 1A imagery. It was for the first time that an attempt was made to identify and map land use/land cover at national level on 1: 250,000 scale for the agricultural year 1988-89 period. In all 26 Centres were involved in mapping process. As a part of this exercise, all the coastal districts of the state were mapped for the spatial extent of land use/land cover wherein areas under built-up land, agricultural land, forest, wasteland and water bodies. In all a total of nearly 19 categories were identified and mapped on this scale. Area statistics for each of the class is computed and tabulated.

FORESTRY

Although the central government in India exercise a control on formulation and regulation of national policies pertaining to forest management, the executive responsibility for managing forests lies with the individual states. Generally forest division, which in most of the cases coincide with the boundaries of district, administrative unit on the basis of revenue returns, is the unit for planning and execution of management activities. The 'Working Plan' is the document prepared for each forest division for managing the forests for a period of ten years, which is subsequently revised.

The wild life management and forest-industry interface management are the other among the important concerns of the forest administration. The management setup *vis-a-vis* problems, the problems of major concern to the forest administration are encroachment of forest by other land uses such as agriculture, homesteads, impact of multi-purpose river valley projects, mining etc., and over- exploitation for timber and fuel requirements of the civilization. While the above are eating into the capital forest wealth, grazing in the forest lands, annual recurrence of fire and shifting cultivation are problems related to regeneration and which need *in situ* solutions. The rapidly shrinking forest cover warrants restocking of degraded forest areas and greening of wastelands.

The operational use of remote sensing technology begins with identification and delineation of forest cover types, to the extent possible in order to conceptualize the ecological status of the forests. A natural sequel to this was to sharpen the experiences and to improve on existing techniques. This involves development of methodologies for extracting information useful for other forestry and ecological studies, and extraction of quantitative information. This information will in turn be useful for use of forest resources protection and conservation, forest ecosystem studies and studies pertaining to forest's role in climate using satellite remote sensing also have been initiated in the country.

The initial survey was done for the entire country using Landsat MSS false color imagery for periods 1972-75 and 1980-92 delineating gross, open and dense forest and prepared 1:1 million scale maps for all the States and Union Territories of India.

The important finding of the study i.e., the core forests (dense) of the country has dwindled from about 14% to 11% of the total geographical area of the country was quite revealing. The impact of this exercise had been tremendous. Firstly it gave a big boost to the awareness of the technique of mapping and monitoring gross forest resources using spaceborne remote sensing techniques. Secondly, it brought out a more defined picture about the rate at which the forest cover was undergoing transformation and finally altogether vanishing. Third it demonstrated that the assessment of gross forest cover of the entire country could be done using this technique in as short a time as three months, involving as minimum manpower as four interpreters and with a cost as low as about half a million rupees (NRSA, 1983). This focussed the attention of all those concerned with resources management to the significance. These time series maps are useful in many ways for resource specialists, including socio-economic studies.

Several studies in the area of technique improvement and methodology development using digital and visual interpretation of data collected from various satellites pertaining to forest type mapping, density and stock mapping, monitoring forest vegetation cover status, encroachments, mapping forest blanks, assessment of production potentials, wildlife studies etc. These were conducted at the various centres of Department of Space and Forest Administration in the country. All of them have important bearing on the forest management. But some of them have even greater significance as they address problems of immediate nature and high priority such as working plan preparation, wildlife management and impact of developmental projects on the forests.

In a recent study the rate of deforestation from 1972-75 to 1980-82 obtained from the study done by NRSA (1983) was correlated to the growth in population in order to evolve a predictive model for forecasting the forest cover according to the increase in population in future.

One of the important ongoing tasks, is the development of a methodology for mapping the grass lands/grazing lands of the entire country. The grazing land of India is in different settings and associations in different regions of the country. Therefore, to evolve an operational methodology it is necessary to evaluate the various analysis and interpretation techniques suited to each region

One of the major pressures to which the forests are subjected to is fuelwood extraction by the rural population. There are two important aspects to be considered in this context. One is that the fuel wood extraction should not exceed the amount that could be supplied on a sustainable basis by forests, which in turn is linked with the annual increment. The other aspect is that the alternate fuel sources should be affordable to the villagers. Agricultural and animal waste has other uses like for fodder and manure. The villagers generally fall into the low-income group and there is a limit for going for other commercial forms of fuel unless with heavy subsidy from government. In order to develop a methodology to assess whether a region is fuel efficient or deficient and understand the fuel energy balance, a pilot study was conducted recently at NRSA. This pilot study revealed that about 67% of the annual increment in forest and other natural vegetation are extracted as firewood alone. This is a rather high value and the timber extracted for other purposes added to it is definitely eating into the capital. The study recommends measures such as better extraction methods and end use, energy plantations etc.

Remote sensing could be introduced into the study of the forest ecosystem dynamics also. This means the study of the dynamics of energy cycles, hydrological cycles and nutrient cycles in the various vegetation and surface cover types of a forest ecosystem and put them into productivity models. This may also contribute to the understanding of the influence of changes in forest ecosystem on the physical systems, which control climate. Here again a multiphase approach could be adopted. The satellite data of the study area could be stratified to delineate the biomass regimes, moisture regimes and temperature regimes.

RIVER CHANNEL MIGRATION AND FLOOD MANAGEMENT

A river/stream in any terrain will be in a state of equilibrium if it has constant discharge, sediment load, slopes Etc. Any change in the controlling variables (which disturb the state of equilibrium), result in aggradation or degradation of the river/ stream course.

Understanding the migration pattern of the river through conventional approach is uneconomical and time consuming. Remote Sensing is a unique technique, which provides information not only of the present river course but also its past courses. Using IRS data, many such river migration areas have been delineated in the Ganga and the Brahmaputra river basins.

Floods and drought affect vast areas of the country, transcending state boundaries. A third of the country is drought-prone. Floods affect an average area of around 9 million hectares per year and 40 m ha area is flood prone. Since complete protection from floods is neither technically for economically feasible, flood management essentially consists of minimizing the damage caused by floods and protecting as large an area as economically justifiable. This can take the following forms:

- flood protection works such as embankments,
- flood hazard zoning to control human activities,
- flood forecasting and warning to enable timely evacuation of people and livestock, and
- Take measures for strengthening flood protection works.

Near real-time flood mapping and monitoring of all major flood events of the country have been carried out on an operational basis since 1986 using satellite remote sensing techniques. Flood inundated areas and damage assessment of Godavari River in the districts of East and West Godavari was attempted by NRSA.

Near real-time satellite monitoring of flood events in Brahmaputra, Ganga, Kosi, Jhelum and Godavari has demonstrated that valuable information can be provided regarding flood affected areas. This helps in planning timely flood relief activities, status of flood protection works requiring reconstruction and strengthening, and flood plain zones subjected to different flood magnitude and frequency. Multiyear satellite data has been used to study river migration and to predict potential areas of flood damage. Flood damage assessment has lead to appropriate allocation of funds for relief measures. ERS-1 SAR data has helped in mapping flood inundation even during cloud covered periods of Godavari floods in 1992. Effective flood hazard zoning will require use of topographic information and discharge/flood records along with multiyear satellite data.

A system is being developed for improved flood forecasting and spatial warning wherein INSAT derived rainfall is incorporated for runoff estimation for better estimation of flood levels. An extensive digital database is being developed in GIS environment to identify the likely impact of flood and to suggest suitable actions to be taken in advance to minimize flood damages. This system is being presently tested in lower Godavari basin. With respect cyclone disaster warning, two warning Centres were established at Bhubaneswar and Visakhapatnam for providing service to maritime states using TV broadcast channels of INSAT satellite. Warning message which originate in these warning Centres are transmitted via satellite to special receiving sets located in cyclone prone areas. By a system of selective addressing specific areas are given warning in local languages. The impact of recent cyclone in November 1996 in Andhra Pradesh was analyzed using IRS 1C, (WiFS & PAN), IRS 1D and Landsat TM data in near real time model (Figure 16.2). These figures clearly depict the eye of the cyclone storm, pre and post cyclone situations.

ENVIRONMENTAL APPLICATIONS

Tackling of environmental problems is intimately linked with many economic and social factors. The environmental consequences of human action involve complex linkages among varied components of earth's environment. Still there is still a greater need for better scientific understanding of the processes involved and the measures human kind should take to preserve the ecological balance while pursuing the some of the urgent development needs. Space technology offers a powerful and speedy means for tackling some of the urgent and large-scale problems related to the environment, as exemplified by many applications realized already in our country. Greater interdepartmental co-operation and sharing of knowledge and experiences play a crucial role in promoting the use of technology, which shows a great promise for assessing and combating environmental degradation.

According to an estimate Andhra Pradesh has 63962 hectares of brackish water area, of which 17,000 hectares have been found to be suitable for prawn culture. No systematic study was so far made to know the area under prawn culture using reliable database.

Mapping and monitoring of prawn cultivation areas in parts of Guntur district over the period 1973 – 1992 was made using multi date satellite data. Figure 16.3 clearly shows the multi-temporal increment in the prawn cultivation areas from 1973 onwards till 1992. The results have shown that prawn culture was negligible in 1973, whereas it was observed in modest levels in 1885 and the area under prawn cultivation was increased significantly during 1990-92. Although initially the prawn culture started in low productive, barren, saline lands because of commercialization in the recent past, fertile areas under paddy cultivation and mangroves were also converted into prawn culture ponds.

Realising the importance of humic substances in production ecology of coastal aquatic environment, a study on characterization of humic acids using IRS 1A data to assess the role of humus in a synoptic manner has been carried out. IT was possible to delineate the role of humic substances and other vital ecological parameters in the coastal primary productivity. Extensive fieldwork was coupled with satellite data analysis for the collection of several ecological parameters including humic substances in various strata of aquatic environment. Spectral models for the individual parameters were developed and a weighted scheme for combining the information was evolved. Coastal aquaculture environments, which are involved in dynamic activity of humic substances, were delineated.

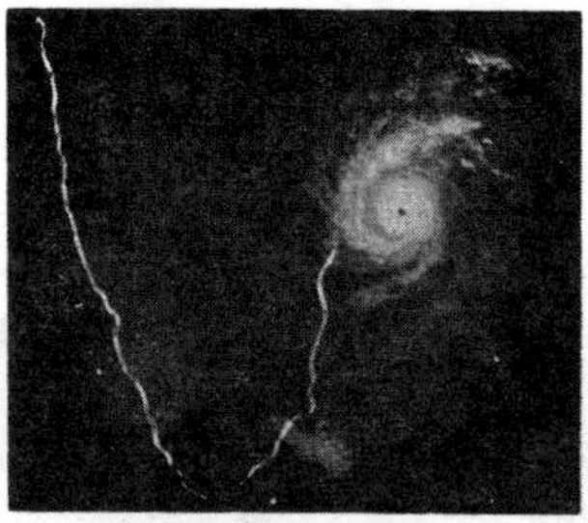

A. NOAA image of cyclone showing eye of the storm

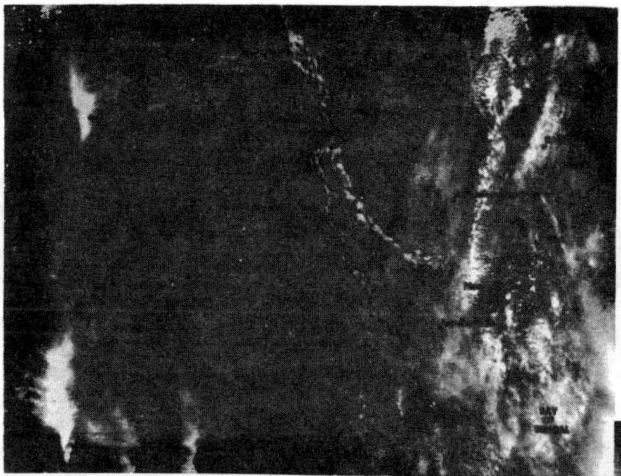

B IRS 1C WiFS data of October, 1996 showing pre-flood situation

C IRS 1C WiFS data of November, 1996 showing post-flood situation

D IRS 1C PAN data of November, 1996 showing cyclone impact in part of Kakinada and environs

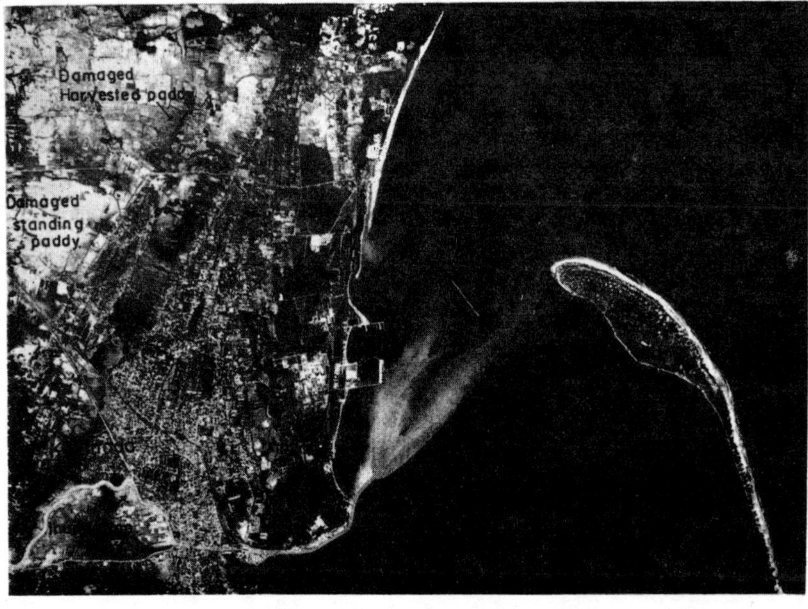

Figure 16.2: Satellite Pictures of Cyclonic Storm in Andhra Pradesh Coast during 1996

Figure 16.3: Computer Classified Prawn Cultivation Areas of Part of Guntur District, Andhra Pradesh

INTEGRATED STUDIES

Primary production activities aimed at higher level of production than the optimum potential of the land meet with frequent failures leading to low production and land degradation in the long run. There is an urgent need to reverse such trends by bringing back the land to appropriate land use practices. The Integrated Land and Water Management Study for Sustainable Development aims at increasing production from and productivity of natural resources especially land and water on a sustainable basis while maintaining ecological balance. Sustainable development of natural resources relies on maintaining the fragile balance between productivity functions and conservation practices through precise identification and systematic monitoring of problem areas in various resources and developmental sectors. Further, it calls for application of alternate agricultural practices, crop rotation, use of bio-fertilizers, energy-efficient farming methods and reclamation of under utilized land and waste lands, planned exploitation of mineral and ground water resources, etc. Optimal exploitation of the resources (both renewable and non-renewable) with proper enriching mechanism calls for cutting across the narrow confines of sectoral approaches, taking a holistic view of the region.

As a first step, natural resources such as soils, groundwater, surface water and natural vegetation are studied and terrain analysis is made for landform and land use using space technology, in addition to studies on slope, watershed and drainage. Three season satellite data from Indian Remote Sensing Satellites are the primary source of data which are studied along with the topographical and other ancillary data and are supplemented with adequate ground data for preparation of natural resources and land use maps. For example several soil profiles in every stratum of landform are studied in the field and samples analyzed in the laboratory to classify the soils at series / association of series level. All the maps were prepared on 1:50,000 scale using Survey of India topographical maps as the base. Climatic factors like rainfall, temperature, humidity, wind speed etc., are also collected and analyzed. All the above parameters are integrated to evaluate the production potential of the land keeping in view the contemporary advancements in the technological development for primary production activities such as agriculture, horticulture, fodder and pasture development, timber production, pisciculture etc. Integrated farming systems such as food crops - horticulture - fodder / pasture or food crops - vegetables - fodder/pasture - pisciculture or horticulture - vegetables fodder/ pasture - forest plantation with efficient irrigation and water management practices and soil and water conservation measures are emerging as appropriate landuse practices for sustainable production and productivity. Geographic Information System (GIS) has been used for integration of spatial data on various resource themes. Suitability of various combinations of land parameters such as soils, ground water quality and potential, slope, landform, landuse / land cover etc., have been linked to primary production activities through rule based decision capabilities of GIS Package. On the basis of this study, alternate development plan showing site-specific primary production activities is prepared.

Implementation of the action plans is envisaged by district authorities under the guidance of the state government and with the active involvement of voluntary agencies. Expert Committees constituted especially for each state / district reviews the action plans before they are taken up for implementation. The Mission Director from Department of Space and Project Directors for each state as well as from Department of Space has been identified for execution of the study. Close linkages with the central agencies viz. Department of Science and Technology (Natural Resources Data Management System), National Informatic Centre, Central Ground Water Board, Central Water Commission, National Water Development Board, Forest Survey of India, Indian Council of Agricultural Research, National Bureau of Soil Survey and Land Use Planning, All India Soil and Land Use Survey, National Bank for Agriculture and Rural Development etc. are established. In addition, numerous voluntary agencies (with expertise at village/ watershed level resource management) are involved in the implementation process. The districts of Srikakulam, East Godavari and Prakasam are among the 174 districts identified in this study.

CONCLUSIONS

The potential of satellite remote sensing in natural resource mapping, monitoring and management is unlimited. Only a handful of examples could be depicted here for the sake discussion. From the preceding discussions, it is clear that it can contribute in various fields ranging widely from geology to marine and offshore applications. As far as coastal zone resource management is concerned, it is influenced mainly by hydrologic phenomena. Thus, it is imperative to lay stress on the behavior of coastal land in a holistic manner taking land – water interactions into consideration. The influences on the environment of coastal biosphere are multi-directional. Ever increasing human needs are changing the maritime environment globally as well as locally. Many a time, the impacts are often adverse. In order to arrest the environmental degradation, it is essential to know the origin and magnitude of the damage in near real-time. But for satellite remote sensing technology, no other conventional source has such proven track record to offer latest information at desired accuracy in a cost-effective manner. Thus, its role is increasingly becoming indispensable.

ACKNOWLEDGEMENTS

The authors are thankful to Dr. D P Rao, Director, NRSA and Sri S K Bhan, Deputy Director (Applications), NRSA for kindly allowing us to use NRSA facilities. The authors are grateful to Dr. A Bhattacharya, Dr. L Venkataratnam, Dr. CBS Dutt, Dr. ST Chari, Dr. NVM Unni, Dr. AN Nath, Sri G Behera, Dr. RS Dwivedi and Dr. PR Reddy for providing us valuable chronicled literature. We also acknowledge the cooperation rendered by the scientific staff of Land Use Division and also the secretarial support of Mr. AV Balasubramaniam.

REFERENCES

Anonymous (1990). Status Report on Crop Acreage and Production Estimation. RSAM/SAC/CAPE/SR/25/90, Space Applications Centre, Ahmedabad, 253.

Balakrishna, P., 1986, Issues in Water Resources Development and Management and the role of Remote Sensing, Technical Report,Indian Space Research Organisation, Bangalore, India.

Bhattacharya, A.K., and Reddy, P.R., 1991, "Hydro geomorphological Mapping for Ground Water Projects in India using IRS imagery", Applications of Remote Sensing in Asia and Oceania, Asian Association of Remote Sensing.

Bowonder, B. Prasad,S.S.R and Madhavan Unni,N.V. 1987: Afforestation in India-Policy and Strategy Reforms. Land use Policy Vol-4, No.7 p-133-146.

Bowonder,B. Prasad,S.S.R and Madhavan Unni,N.V., 1987a: Deforestation and Urban centres in India. Environmental Conservation. Vol.14, No.1, p-23-28.

Bowonder,B., Prasad,S.S.R and Madhavan Unni,N.V., 1986: Fuelwood prices in India-Policy Implications, Natural Resources Forum, Vol.10, p-5-6.

Central Water Commission, 1988, Water Resources of India, Publication No.30/88, New Delhi.

Dwivedi, R.S., and Singh, A.N., 1983, "Soil Resource Inventory of Ghataprabha Left Bank Canal Command Area - Part of Bijapur and Belgaum Districts of Karnataka for Water Management Planning using Landsat Images, - National Symposium on Remote Sensing in Development and Management of Water Resources, India.

Deekshatulu,B.L., Madhavan Unni,N.V., Murthy Naidu,K.S. and Chennaiah, 1989: Conf. Ind.Sci. Acad.Bhopal, on environ and Forest degradation.

Dutt,C.B.S., Ranganath,Y.V.S., Parthasarathi and Lahan, P,1988: Application of high resolution IRS-1A, LISS-II data for vegetation cover mapping, Proc. of IRS Seminar, Hyderabad (in press).

Forest Survey of India, 1991: The State of Forest Report. Min. of Environment and Forests. Publication. p-7-50.

Gautam, NC , Raghavswamy, V, and Nagaraja, R , 1994: Space Technology and Geography, Published by National Remote Sensing Agency, Hyderabad.

Kumar, S., Haefner, H., and Seidel K., 1991, Satellite Snow cover Mapping and Snowmelt runoff Modeling in Beas Basin " IAHS Publication, P 101-105.

Kushwaha,S.P.S and Madhavan Unni,N.V., 1986: Application of remote sensing techniques in forest cover monitoring and habitat evaluation - A case study at Kaziranga national park, Assam. Proc. Sem. cum Workshop on Wildlife habitat evaluation using remote sensing techniques, Oct 22-23 1986, Dehra Dun. p-238.

Madhavan Unni, N.V. and Murthy Naidu, K.S. 1989: Monitoring seasonal and annual changes in environment using satellite remote sensing - A case study in Keoladeo national Park, Bharatpur. Proc. Seminar on Wetland Ecology and Management, Bharatpur, Feb. 21-24.

Madhavan Unni, N.V., Murthy Naidu, K.S., and Kushwaha S.P.S., 1986: Monitoring forest cover using satellite remote sensing techniques with special reference to wildlife sanctuaries and national parks. Proc. of seminar cum workshop on wildlife habitat evaluation using remote sensing techniques. Oct 22 & 23, p-146.

Madhavan Unni, N.V., Roy, P.S. Gautam, N.C., Murthy Naidu, K.S., Shashi Kumar, V., and Singh, A.N., 1985: Application of remote sensing technology to environmental problems - An overview with special reference to India. Proc. of 2nd World Congress on Engineering and Environment. Inst. of Engineers (India), New Delhi.

Madhavan Unni,N.V., 1990: Space and Forest Management in India. Proc. Space & Forest Management, Special Current Event Session, 41st IAF Congress, Dresden, Germany.

Madhavan Unni,N.V., Roy,P.S. and Murthy Naidu, K.S., 1983: Environmental studies through remote sensing and mapping of Idukki area in Kerala, Part.I, NRSA Report.

Madhavan Unni., N.V., Roy, P.S., Jadhav, R.N., Tiwari, A.K., Sudhakar,S, Ranganath, B.K., and Dabral, S.L. 1991, IRS-1A, Application in Forestry Current Science, Vol. 61 Nos. 384, Special Issue.

Mohanty, R.B., Mohapatra, Mishra.D. and Mohapatra, 1986. 'Application of Remote Sensing to Sedimentation Indian in Hirakud Reservoir ' Joint Report of Orissa Remote Sensing Applications Centre, and Hirakud Research Station, Orissa, India.

Moore G.K., 1988, 'The role of Remote Sensing in Ground Water exploration', Proceedings of Joint Indo- US workshop, Hyderabad, India.

Narendra K.,1989 ' The study of Hydrologic Response of Irrigation Tanks in Prakasam District, Andhra Pradesh State using Satellite Data', M.Tech. Thesis, Jawaharlal Nehru Technological University, Hyderabad, India.

NRSA, 1983: Report on forest mapping cover in India from Satellite Imagery 1972-75 and 1980- 82.

NRSA, 1989. Manual of Nationwide Land Use / Land Cover Mapping using Satellite Imagery, Part – I

NRSA, 1990. Project Report. IRS Utiisation Programme: Soil erosion mapping.

NRSA, 1992. Project Report. Acreage estimation of soybean and sunflower crops.

NRSA, 1994. Project Report. IRS Utilisation Programme: Soil resources mapping.

NRSA 1986, 'Ground Water Potential Maps of Drought prone districts of Maharashtra prepared based on visual Interpretation of Landsat Thematic Mapper Data', Project report, National Remote Sensing Agency, Hyderabad, India.

NRSA 1991, 'Major Crop Identification and Estimation of Irrigated area in Nigamsagar command area using Landsat TM and IRS IA LISS I Data', Project Report, National Remote Sensing Agency, Hyderabad, India.

NRSA and APSRAC, 1996 : Andhra Pradesh cyclonic storm damage assessment through IRS 1C. INTERFACE, Vo. 7., No.4.

Parihar, J.S. Kotwal, P.C., Panigrahi, S and Chaturvedi, N. 1986a: Study of wildlife habitat using high resolution space photographs - A case study of Kanha National Park, Report ISRO SP-17-86, p-65-82.

Parihar, J.S., Chaturvedi N., Panigrahi, S and Kotwal, P.C. 1986: Study of network of wildlife reserves in Eastern Madhya Pradesh using remote sensing data. ISRO Report SP-17-86, p-83-94.

Rao D.P., 1988, 'Integrated Drought Management ', Proceedings National Symposium on Remote Sensing for Rural Development, Hissar, PP 387-398.

Rao D.P., 1995. Remote Sensing for Earth Resources, published by Asso. Of Exploration Geophysicists, Hyderabad

Rao, T.H., Rao,C.R., and Vishvanathan,R., 1988, 'Capacity Evaluation of Sriramasagar Reservoir by using Remote Sensing Techniques', Proceedings of 54th R & D session, CBIP, Ranchi,India.

RRSSC, 1991, 'Capacity Evaluation of Ghataprabha Reservoir using Digital Analysis of LISS II Data ' Project Report, RRSSC B 3/91, Regional Remote Sensing Sensing Centre, Bangalore, India.

Sahai B., Kalubarme, M.H. and Jadav, K.L., 1985, 'Ecological Studies in the Ukai Command areas', International Journal of Remote Sensing , 6.

Sharma, R.P. and Sharma, M.K. 1982: Degradation of forest cover of Doon Valley. Proc. of Symp. on resources survey for land use planning and environmental conservation, Dehra Dun, p-13-19.

Shedha, M.D. and Dutt, C.B.S. 1982: Spatial and structural changes in the forest resource of Doon Valley using Remote Sensing Techniques, Proc. of Symp. on resources survey for land use planning and environmental conservation, Dehra Dun, p-24-26.

Singh, J.S. Tiwari, A.K., and Saxena, A.K., 1985: Himalayan forests - A net source of Carbon for the atmosphere. Environmental conservation. Vol. 12, No.1, p-67-69.

Singh, J.S., Uma Pandey and Tiwari, A.K. 1984: Man and Forests - A Central Himalayan case study. Ambio vol. 13, No.2, p-80-87.

Sinha, R.S., Ashok Kumar, and Saxena, R.C., 1990, 'Hydrological Appraisal in part of Sharda Sahayak Command Area of Utter Pradesh - A Study based on Remotely Sensed Data' Proceedings of National Symposium on Remote Sensing for Agricultural Applications, New Delhi, India.

Thiruvengadachari S., 1983, Irrigated Agriculture in Cumbum Valley in South India : A Landsat study', 17th International Symposium Remote Sensing of Environment, Ann Arbor, U.S.A.

Thiruvengadachari S., and Hari Krishan, J., 1992, 'Satellite Monitoring of Irrigation Command Area under Rajolibanda Diversion Scheme Project, Mahaboobnagar,

Andhra Pradesh', Proceedings of National Level Group Discussions on Irrigation Management, Trichy, India.

Thiruvengadachari S., and Manavalam P., 1983, 'Evaluation of the use of Satellite Sensing Techniques for Inventoring Tank Irrigation in Tamil Nadu State', National Seminar on Natural Natural Resources Management System, Hyderabad., India.

Thiruvengadachari, S., 1990, ' Satellite Surveillance for improved country wide monitoring of agricultural Import Conditions', Proceedings of National Symposium on Remote Sensing for Agricultural Applications, New Delhi, India, PP 3&9-408.

Thiruvengadachari, S., and Jonna S., 1992 'Irrigated Command area Inventory using IRS IA LISS I Data' Proceedings ISY Conference on Remote Sensing and GIS, JNT University Hyderabad, India.

Tiwari, A.K. and Kudrat, M 1988: Analysis of vegetation in Rajaji National Park using IRS data. Proc. of IRS Seminar, Hyderabad. Dec. 21-23 (in press).

Tiwari, A.K., Saxena, A.K. and Singh, J.S. 1985: Inventory of forest biomass for Indian Central Himalayas. In Environmental regeneration in Himalayas - Concepts and Strategies, Gyanodaya Prakashan, Ed:J.S.Singh

Ventataratnam, L. And Thamappa, S.S., 1993 : Mapping and monitoring of areas under prawn farming. INTERFACE, Vol. 4, No.2.

17

COASTAL FISHERIES IN ORISSA: AN IMPORTANT ECONOMIC ACTIVITY

S. Ayyappan and J. K. Jena

INTRODUCTION

Nature has endowed Orissa with a wide variety of fishery resources, both natural and manmade. The marine fishery resources of the state include 480km long coastline with a continental shelf area of 24,000 km^2. The brackishwater resources include over 1,00,000 ha of brackishwater lagoons, 32,587 km^2 of brackishwater area that are suitable for culture fisheries development, besides 8,100 ha brackishwater swamps and tidal mudflats and an estimated 2,97,850 ha of estuaries. The long coastline spreading over six districts, *viz.*, Balasore, Bhadrak, Jagtsinghpur, Kendrapara, Puri and Ganjam and the vast continental shelf account for 6% and 5.3% of the Indian coastline and continental shelf respectively. (Figure 17.1)

The fisheries potential of the exclusive economic zone of the state as estimated by Fisheries Survey of India is 1,69,000 metric tonnes including 100,000 tonnes in continental shelf i.e. upto 200 m depth and 69,000 tonnes in continental slope i.e. beyond 200 m depth. However, Central Marine Fisheries Research Institute estimated it to be as much as 2,08,000 tonnes including 1,80,000 tonnes from upto 50 m depth and rest beyond 50 m.

In contrast, the marine fish landing of the state is mostly restricted within a depth range of only 50 m. The fish production of the state from marine sector is showing a slow but steady increase over the last few years, the details of which are as follows (Table 17.1):

Table 17.1: Marine Sector Output in Orissa

Year	Production (tonnes)
1992-93	1,19,376
1993-94	1,03,925
1994-95	1,22,892
1995-96	1,23,200
1996-97	1,33,462

As regards to the fish catch during 1996-97, the major share of the landings were contributed by sciaenids (12.23%), followed by elasmobranchs (7.13%), catfish (6.12%), hilsa (5.22%), pomfrets (4.38%), other clupeids (4.21%), polynemids (3.80%), and prawns (4.17%) besides the miscellaneous varieties contributing as much as 52.74% of the total production.

The state has witnessed remarkable development in brackish water farming sector too. The development of water area was started only during early eighties, with only 23.5 ha in 1983-84, that went up with an increasing trend till 1991-92, in which year the development was as much as 2075 ha, which, however, showed a decline thereafter. The production of shrimps from these areas showed an increase from only about 12 tonnes during 1983-84 to as high as 6957 tonnes during 1995-96 which showed a marginal decrease to 6627 tonnes in following year. Presently 12,439 ha of area is under shrimp farming including 11,482 ha (92.3%) under extensive, 320 ha (2.52%) under modified extensive and 637 ha (5.13%) under semi-intensive cultivation.

Freshwater fishes such as carps being one of the main ingredients of food of the people in the state, inland fisheries and freshwater aquaculture assume great importance in the socio-economic scenario of Orissa. The freshwater resources comprise 114,125 ha of ponds and tanks; 180,000 ha of lakes, swamps and bheels; 256,000 ha of reservoirs and 155,000 ha of rivers and canals. While freshwater sector is poised for a potential of over 252,000 tonnes, the present level of exploitation (1996-97) is limited to just over 55% of the estimated potential, i.e. 127,293 tonnes. At the same time the exploitation

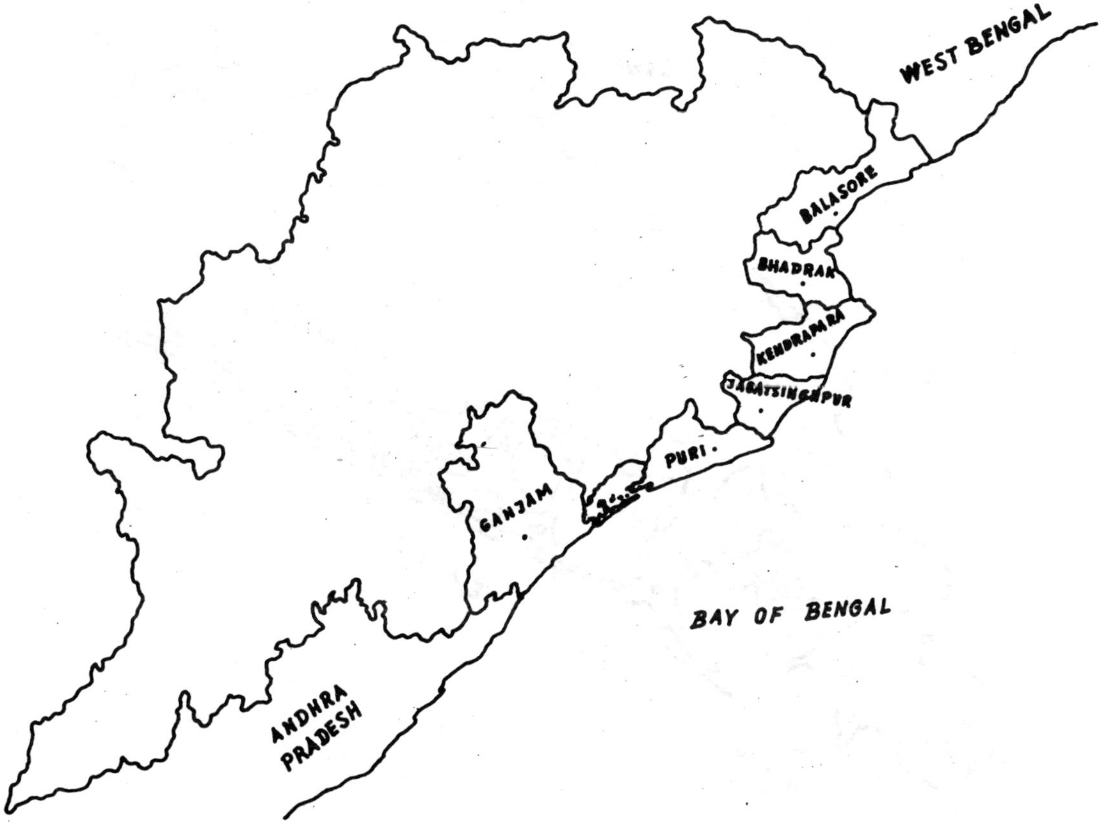

Figure 17.1: Coastal Districts of Orissa

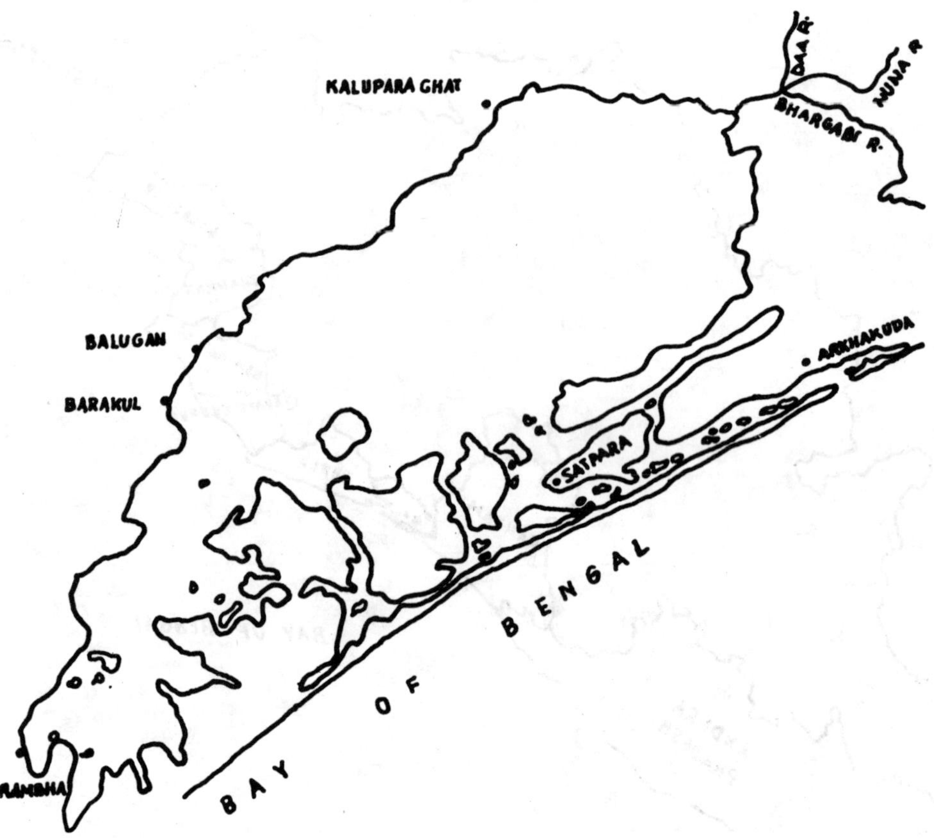

Figure 17.2: **Chilka Lagoon**

from the brackish water sector is just over 11,000 tonnes against the production potential of 65,528 tonnes.

The coast of Orissa is broadly divided into two regions i.e. north Orissa coast comprising the coast of Balasore (80 kms), Bhadrak (50 kms) and Kendrapara (68 kms) and south Orissa coast comprising Jagatsinghpur (67 kms), Puri (155 kms) and Ganjam (60 kms) districts. While the north Orissa coast is shallow, muddy and calm which is characterized by tidal flats and extensive river deltas, the south Orissa coast has the usual surf beaten sandy beaches. The State possesses as many as 30 important rivers with a total length of 4362 kms and most important among them that flow into the Bay of Bengal through the state are Mahanadi, Baitarani, Budhabalanga, Subarnarekha, Devi and Rushikulya. Besides these, the state possesses two important ecosystems of international importance due to the rich biodiversity for flora and fauna, they being Chilka lake and Bhitara Kanika, the details of which are discussed here.

CHILKA LAKE

Chilka lake is one of the biggest brackishwater lagoons in Asia located along the east coast of India between latitudes 19° 28' and 19° 54'N and longitude 82° 05' and 85° 35'E in Puri and Ganjam districts of the state of Orissa (Figure 17.2). The lagoon has a original water spread area of 906 km^2 in the summer and of 1165 km^2 in the rainy season, with an average area of about 1000 km^2. However, at present, the area has been reduced to about 700 km^2 as per the survey conducted by ORSAC. The lake is separated by a sand bar of 60 km in length formed by the waves and wind in the fore-shore area. The sand bar is concave towards the sea and is irregular on the other side having a number of cuspidate spits. The outer channel runs parallel to the lake and sea, and is separated from the latter by a narrow sand bar and from the former by a series of peninsular islands. On its course to south-west, the channel divides into two branches at Satpara. While one branch continues along the original course and eventually merges into a network of swamps and water ways, the other broader branch taking turn at right angles, finally reaches the main part of the lake at a point called Mugger Mukh.The opening of the Mugger mukh remains extremely shallow during the summer months. The channel and its complex physiography permits limited tidal incursion of sea water into the lake, thus keeping the water of the lagoon in brackish for most of the period. The mouth of the Chilka lake appears to be highly variable and shows a gradual shift to the north-eastern side due to the continuous occurance of sand bars towards the north. The lake which was connected by an artificial manmade channel (Palur channel) connecting Rambha bay and Rushikulya estuary has been narrowed off at present due to the heavy siltation and the water holdings along the channel are used for prawn farming. The depth of the lake varies from 2.5 m in summer to 3.6 m in flood season. The lake receives huge amount of flood water from river Daya, Nuna, Ratnachira, Kania, Malgoni, Dhanua and Salia. The various tributaries also carry huge amount of silt (13 million tonnes/year) along with flood water, thereby rendering the lake shallower over the years.

Chilka lake is known to possess a great biodiversity of species of both fishes and shellfishes. A total number of 217 species of fresh/brackish water fish comprising 147 genera, 71 families and 15 orders so far were reported from Chilka lake. As many as 24 species of prawns and shrimps are also reported from the water of Chilka lake, besides 9 families of crabs comprising 28 species. Further, from the studies made so far, a total number of 136 species of molluscs under 66 families were reported from the lake. The fish yield from the lake during 1948-1952 was ranging from 3460-4004 tonnes/yr with an average production of 3716 tonnes/year which reduced to an average production of 3272 tonnes/year for the period 1965-66 to 1969-70. Later, the production showed to be increasing in subsequent years, reaching maximum of 8872 tonnes in the year 1986-87, which however, showed a decreasing trend thereafter. The annual fish production of the lake in subsequent years are given in Table 17.2 below:

Table 17.2: Annual Fish Production in Chilka Lake

	Annual landings from Chilka		
	Fish	Prawn	Total (M.T.)
1987-88	6863	1241	8104
1988-89	5211	917	6128
1989-90	5493	1177	6670
1990-91	3792	481	4273
1991-92	3680	876	4556
1992-93	3207	951	4158
1993-94	2799	686	3485
1994-95	1239	176	1415
1995-96	1056	213	1269
1996-97	1352	281	1633

The reasons for such reduction in total production of fish and prawn from the lake are attributed to over-exploitation, lack of conservation measures and reduction in recruitment over the years. The other reasons affecting the recruitment and production of different economical varieties are:

(i) heavy siltation at the bar mouth restricting the ingress of saline water into the lake and vice-versa, thereby limiting the migration of several commercially important species for their breeding and nursery phase,

(ii) reduction of depth of the lake over the years due to the heavy deposition of silts carried through river drainage during monsoon,

(iii) drastic changes in hydro-biological parameters due to narrowing of bar mouth and disposal of enormous amount of wastes into the water body and

(iv) encroachment by the shrimp farmers by making barricade, excessive infestations with many aquatic weeds of freshwater forms and seaweed.

The fish production of the lake is mainly contributed by a few commercially important fish species, viz., Mugil cephalus, Liza spp., Pseudosciaena coiber, Polynemus tetradactylus, Nematolosa nasus, Mystus gulio, Hilsa ilisha, Gerres setifer, etc. and shrimp species, Penaeus indius, P. semisulcatus, Metapenaeus monoceros, M. dobsonii, etc. Further, among the crab species Scylla serrata and Neptunus pelagicus also form important fisheries in certain parts of the lake.

A multi-disciplinary expedition study was conducted, spreading over three calendar years during 1985-87, with 219 sampling stations evenly distributed all over the lake and outer channel during three major season, viz., summer, rainy and winter. The lake, declared as a "Wetland of international importance" under Ramsar convention in 1979 and later ratified by United Nations Conference on Human Environment at Stockholm, has been the cause of appreciable concern of everyone due to the imminent environmental threats to the ecology and fishery resulting from several anthropogenic activities.

The faunal diversity of the lake is impressive by all accounts. Besides the rich diversity of finfishes and shellfishes, the lake also possesses over 150 species of avifauna, two species of crobs dolphins besides several migratory birds. Further, the lake is heavily infested with aquatic weeds of different kinds. The central and southern sector are blanketed by seaweed species like Gracilaria lichenoides, G. confervoides and

Potamogeton pectinatus. The other macrophytes of importance are Najas falcioulata send Halphila ovata besides heavy infestation of water hyachinth (Eichhornia crassipes) at the peripheral edges.

BHITARA KANIKA

Bhitara Kanika is a unique mangrove ecosystem with varied species of flora and fauna and is the second biggest mangrove forest in the country next to Sunderbans of West Bengal (Figure 17.3). Located at Kendrapara district of Orissa, it supports varieties of common and endangered wild life and rightly notified as a sanctuary in 1975. The area of the sanctuary spreads over 650 km^2 having a forest cover of 380 km^2, out of which the mangrove forest spreads as much as 115.5 km^2. This unique habitat possesses as many as 62 mangrove species, out of the total number of 67 species recorded in the country. The mangrove also possesses some rare mangrove plants which are not reported from any other place of the country *viz., Heritiera kanikensis* and *Merope angulata*. Thus, it can be considered as a true representative of the mangrove regions of the country, despite the comparatively low spread of the area. Thick mangrove forests, the creek and the beach provide an ideal habitat for wild life, both aquatic and terrestrial. The mangrove is considered the natural home for many reptiles and amphibians, most important among them being the estuarine crocodile, *Crocodilus porosus*. The mangrove habitat also act as a nursery ground for many fish and shellfish species of commercial importance.

Gahirmatha beach of Bhitara Kanika is the biggest nesting ground of Olive Ridley sea turtle (*Lepidochelys olivacea*) in the world. The beach of Gahirmatha with a stretch of 35 kms witnesses the most unique event of the arrival of half a million female olive ridley every year during late December to January and again from mid March to April. The turtles nest both in east coast as well as west coast of India but the rookery at Gahirmatha and another near Devi river estuary together host one of the largest aggregations of olive ridleys in the world. These two sites receive about 500,000 olive ridley nestings annually and the turtles lay about 500 million eggs, considering each one laying atleast 100 eggs at a time. The mating and nesting events occur at predictable times every year with minor fluctuations due to prevailing local environmental conditions. The mass nesting usually occurs in the season when the average width of the beach is at its maximum providing maximum space to the female for nesting and when the beach experiences a particular temperature range necessary for the normal and successful development of embryos for maximising the hatching rate. Though the number

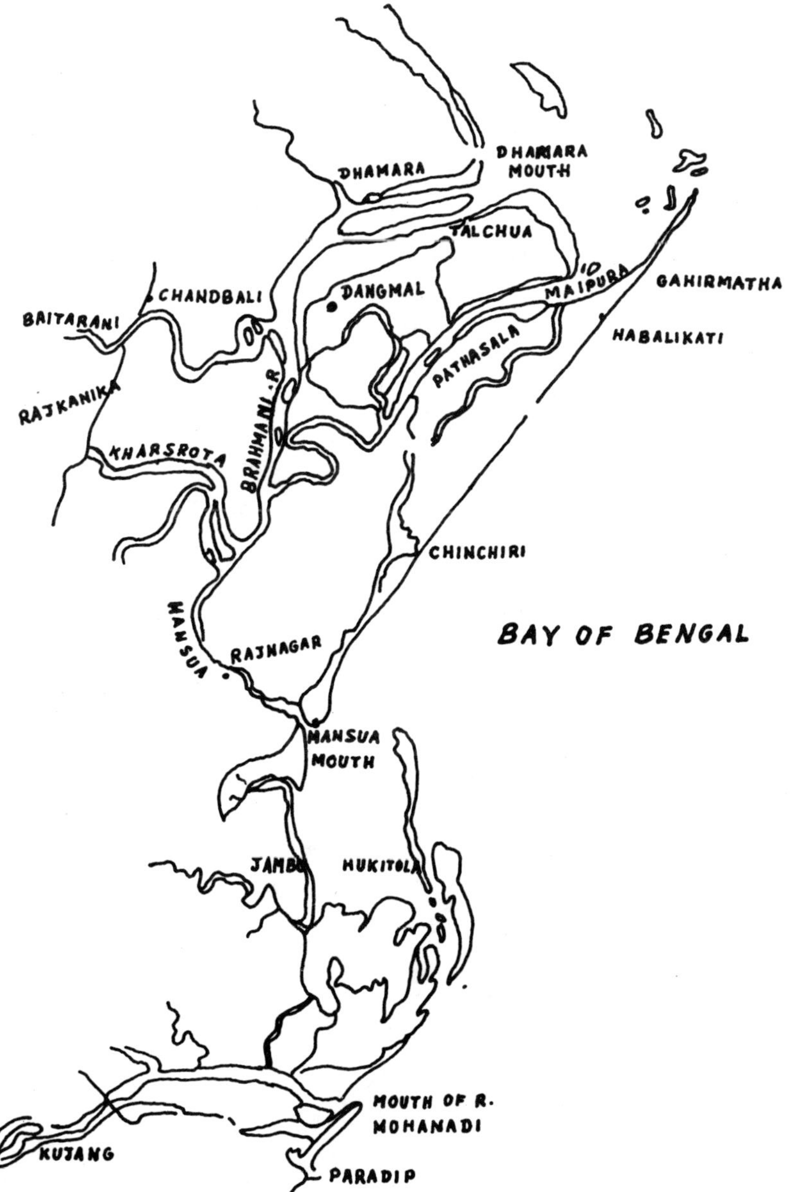

Figure 17.3: Bhitara Kinika

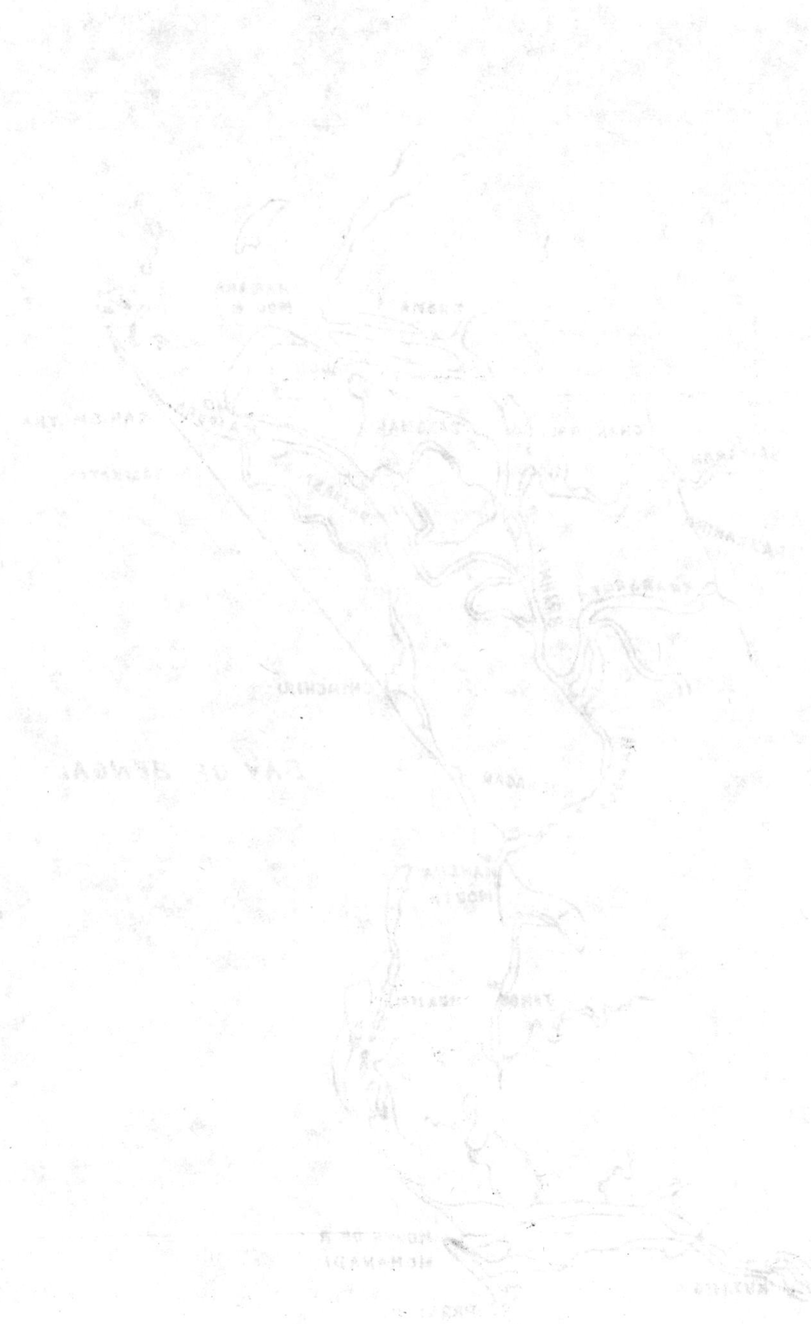

of adult female turtles that visit the beach varies depending on the above mentioned criteria, yet they are in hundreds of thousand.

The nesting beach along this stretch is backed by dense mangrove forests of *Avicennia marina, A. officinalis, A. alba, Rhizophora mucronata, Ceriops tagal, Aegialitis rotunidfolia* and many other species. This mangrove forest is the home of notorious egg predators of sea turtles such as wild pigs, jackals and water monitors. Each adult turtle weighs 30-50 kg and thus some thousands of tonnes of these turtles that come to the beach make the beach an unique one. However, during recent years the population of these sea turtles have been depleted primarily because of human consumption of eggs in very large scale from nesting beaches, consumption of eggs by non-human predators like wild boars, bags, birds, etc., poaching of adults in large scale during the breeding season, trade in turtle products like leather, hide and turtle shell, and disruption, alteration and erosion of nesting beaches.

Besides olive ridley sea turtles and brackish water crocodiles, the environment also provides shelter for many other important animals like spotted deer, sambar, leopard, fishing cat, porcupine, otters, etc. and reptiles like python, king cobra, kraites, water monitor lizards, etc. The sanctuary also forms nesting sites of over 170 varieties of birds including different species of king fishers, stork, heron, egret, white bellied sea eagles, kite, common snipes, curlews, sand pipers, cormorants, darters, white breasted water hens, lapwings, parakee, myna, Gray hornbills, red jungle fowl, sea gulls, etc. including many migratory forms coming from north Asian region.

CONCLUSIONS

As in any other coastal state, Orissa too has a great influence of the seas and beaches on the lives of the people. Apart from fisheries, the resource provides for tourism in a large measure. Consequently, associated infrastructure facilities have been added along the coast line. These activity in the coastal area need to be coordinated and regulated for achieving sustainable coastal resource management. While fishing acitivity needs to be enhanced to realise the potentials of the nearshore seas in the state along with post-harvest processing facilities, greater attention is needed towards environmental conservation of the coastal ecosystem. In this context, the proposal for a comprehensive coastal environmental management plan with scope for conservation of biodiversity of rich flora and fauna with due emphasis on sustainable development, needs to be formulated.

18

COASTAL POLLUTION IN WEST BENGAL: AN EMERGING PROBLEM

Anil Baran Bhunia, Tapas Kr. Gupta and Debanjan Gupta

INTRODUCTION

The coast is a transitional zone between the land and the sea where both the environments interact intimately with each other and hold highly diverse and dynamic ecosystems. The coastal zone is generally viewed as common resource available to all. Uses of this space are multiple and include human settlement, food production, derivation of energy, minerals and other raw materials, tourism, recreation, transport and development of industries including ports and harbours.

Nearly 60% of the world population (about 3 billion) live within 60 km of the coastline. Two thirds of the cities of the world, with population over 2.5 million each, are situated near coasts. With the growing trend of population migration into the coastal zones, the activities on the coast and utilisation of resources are on the increase. The UNCED Rio '92 conference has already focussed the problems associated with the coastal environment and its resources. Apart from natural hazards threats posed by sedimentation, erosion, pollution, unregulated fishing practices, harmful algal bloom in coastal waters have been major problems world-wide in the present times.

To protect the aquatic and marine life some of the programmes had been undertaken all over the world to study the different aspects of coastal zone. Some of these programmes are listed in Table 18.1.

Table 18.1: Important Programmes on Coastal Issues

Programme	Executing Agency
Coastal Marine systems	UNESCO project
Regional Seas Programme	UNEP
Fisheries programme	FAO
Marine Meteorological programme	MMO
Integrated Global Ocean Station System	IOC and WMO
Coastal Zone Activities	IUCN

In India, organised efforts were initiated since 1986 to assess the state and trend of marine coastal pollution in the seas around the country upto 5 km offshore through systematic monitoring under the project 'Monitoring of Indian Coastal Waters'. The project was administered jointly by the Dept. of Ocean Development (DOD), the Central Pollution Control Board (CPCB) and the National Institute of Oceanography (NIO). Subsequently, the DOD has launched a project 'Coastal Ocean Monitoring and Prediction System (COMAPS) in 1990 extending the study area upto 25 km offshore through ten executing units in the country.

WEST BENGAL COAST

West Bengal Coast, by virtue of having country's largest river outfall- the Hooghly (Ganga) and most critical habitat - the Sundarban mangrove forest, has occupied an important place in world map. The coast is about 200 km long stretching from inter State boundary with Orissa in Midinapore district to the international boundary between India and Bangladesh. Along the strech there are important beach resorts at Digha, Bakkhali, places of important pilgrimmage like Ganga Sagar in Sagar Island, Birds Sanctuary and unique tiger habitat in Sundarban forests. The Hooghly-Matla river system including other estuaries namely, Bidya, Thakuran, Saptamukhi with their innumerable tidal creeks and channels form an extensive network in eastern sector of the coast.

Being well recognised in the World Heritage list the Sundarbans in the territory of India and Bangladesh constitute about 12 percent of total world mangrove resources. The Ganga-Brahmaputra river system reportedly brings about one-fifth of total sediment load (i.e. 2000 million tonnes per year) transported annually by all rivers of the world to the seas and thus created the vast deltaic plain. Out of 9,630 km2 total area of the Sundarbans about 4264 km2 area is in the West Bengal which includes 35 percent water area.

Sundarban Mangrove forest contains more than 60 percent of India's total mangrove forest resource with not less than 70 species and has developed into world's unique mangrove tiger land ecosystem which is also a Biosphere Reserve. Besides a number of forest products like timber, fire wood, honey etc. this unique habitat holds myriad of flora and fauna of which most importantly are more than 100 species of fishes.

MAJOR COASTAL ACTIVITIES

Resources in both land and water fronts are intrinsically linked with the development of state's economy. Besides fishing, aquaculture, salt production, transport and use of mangrove forest products the coast holds tourist resorts having unique scope for future potential as also development of ports and harbours for enhancement of trade and commerce in the state.

Sources of Pollution

As such the state is very rich in industrial settlement. But the coastal belt is almost devoid of any industry. However, at about 20 nautical miles land ward from the coast there is an industrial complex at Haldia. Most importantly, the major number of industries in Howrah, Hooghly, Durgapur and Calcutta are located by the side of the Hooghly and Damodar the effluent of which ultimately find way into the Bay of Bengal. In total, 62 numbers of grossly polluting industries (State Board's Survey Report 1997) situated along this river system are the major sources of industrial wastes which finally transported into the coastal areas. Besides, the sewage of the Calcutta Metropolitan area is transported to the coastal areas mostly through the Matla river in the east and partly through the Hooghly in the west. In addition, the waste generated in port related operations at Haldia and Calcutta and fishing harbour at Sankarpur, oil spillage and bilge water discharge from ships as and also rejects of coastal aquaculture firms are becoming the increasing threats to the coastal water quality.

Another major pollution generation source is the region near Digha which is a major tourist spot in West Bengal. There are about 400 hotels, holiday homes, some eating joints and sweet meet shops. The waste generated from these sources are directly discharged into the coastal waters.

Coastal Water Quality

Being characteristically shallow coastal configuration the watermass remains under steady mixing process due to high tidal amplitude, turbulance and strong monsoonal gyre often accompanied by cyclone-storms. Results of coastal water quality monitoring during the past few years (Reports of Central Pollution Control Board) revealed that the water quality is good interms of physico-chemical determinates.

As the zone is under tremendous influence of river Hooghly, the salinity showed mesohaline condition throughout the years with the range generally falling between 4 and 27 ppt. Dissolved oxygen was also very consistent throughout the years and generally varied between 5 and 7 mg/l. Nutrients particularly inorganic phosphate and nitrate concentration showed some minor fluctuations through seasons. In general, higher concentration prevailed during monsoon and post monsoon seasons than in the winter. Though substantial concentration of nutrients are available throughout the year, the utilization/uptake of the same by the microbiota (phytoplankton) seemed to be less because of high suspension load interferring the system as evident from low range of primary production. Heavy metals viz. lead, mercury and cadmium concentration were found to be less both in water and sediment. High concentration of total coliform bacteria in the near shore water and the presence of pathogens on some occasions have been important concern for safeguarding the water quality from use viewpoint.

Major Problems and Issues

Some of the major issues distrubing the coasatl ecosystem of West Bengal are listed as folows:

- Erosion of river banks, islands, siltation/sedimentation in the river mouths-causing navigational hazards (Hooghly mouth).
- Subsidence/destabilisation of beaches (Digha).
- Exploitation/felling of mangrove trees causing erosion of islands (Sundarbans).
- Unregulated fishing/over fishing (loss of Biodiversity)
- Unregulated abstraction of ground water in coastal zone
- Accidental oil spills and discharge ships waste.
- Growing microbial pollution (including pathogens) in coastal waters.

POLLUTION ABATEMENT MEASURES

The Pollution Control Boards at central and littoral states have adopted 5 km into the sea from the highest hightide line as an extent, over which their authority to control waste discharges will extend for the purpose of the Water (Prevention and Control of Water Pollution) Act, 1974. Taking the matter into the task, the State Board has already issued directives against some of the hotels to install water and air pollution control systems. Construction of one Common Effluent Treatment Plant for the hotels and other eating joints is being planned for treatment of all liquid waste generated from these sources. Also the successful run of sewage treatment plants under the Ganga Action Plan of the Central Government may well restrict the raw sewage discharge into the waterbodies and thereby contributing substantial reduction on bacterial pollution in Hooghly as well as coastal waters.

With the increasing social need, shipping/water ways transportation of different commodities would increase and thus the number as well as the movement of ships and small crafts all would increase with the passage of time. In case of accidental spillage of oil/harmful substance, an action plan including immediate reporting and effective response to containment and treatment is necessary. For this purpose, implementation of an effective contingent plan jointly by the shipping authority and oil companies may be of great rescue to the sensitive ecosystem of West Bengal including important estuarine zone of Hooghly (Ganga).

Under the Environment (Protection) Act, 1986, rules were framed for regulation of activities in the coastal area and declaring coastal states as Coastal Regulation Zone (CRZ) and regulating activities in the CRZ. Accordingly, three zones have been demarcated (zone I,II,III as per MoEF Notification 1991) for the purpose of regulating different activities so as to ensure restoration and development of the coastal zone environment in a sustainable basis.

REFERENCES

Monitoring of Indian Coastal Waters- Summary Report *by Central Pollution Control Board, New Delhi.*

Pollution Potential of Industries in Coastal Areas of India, *Central Pollution Control Board, New Delhi.*

Coastal Pollution, *Central Pollution Control Board, New Delhi.*

Triennial Report on Coastal Ocean Monitoring Along the North-East Coast of India, Department of Ocean Development Project (1991-1993) *by East-North East Regional Office and Department of Ocean Development Unit, CPCB.*

19

LAND USE PATTERN WITHIN THE COASTAL REGULATION ZONE OF WEST BENGAL : APPLICATION OF SPOT 1 HRV MLA DATA

S. Bhattacharyya, J. Nanda and S. Nayak

ABSTRACT

With the present day concept of sustainable utilisation, specialists as well as the policy makers are interested in the wise use of the coastal resources. To enable it, Government of India has imposed certain regulations and has declared a particular portion of coastal regions as a regulatory zone, termed as Coastal Regulation Zone(CRZ). Idea is to develop the coastal sector on the basis of certain management plan for sustainable development and to protect the coastal environment. The management plan needs to be prepared on the basis of objective situation of this zone and with an holistic approach. In order to prepare a baseline information of CRZ of West Bengal, the present study of landuse pattern of this zone was carried out on 1:25,000 scale utilising the Remote Sensing technique and based on a classification system with wide applicability. The maps were finalised after detail ground truth checking and the individual landuse units were measured. This chapter is likely to throw certain light on this particular zone of West Bengal not studied heretofore and would serve a baseline for preparation of Coastal Zone Management Plan for the State of West Bengal.

INTRODUCTION

The Coastal Zone environment, a unique geological-biological-ecological domain, is the most dynamic of all the known environments. It is also unique in terms of productivity. In true sense, coastal resources are varied and diverse and can be classified under different units like land, forests, coastal waters (including enclosed and semi enclosed seas, estuaries, inland waters), minerals and hydrocarbons, living marine resources etc. These are regarded as renewable or semi-renewable natural resources.

The Coastal Zone in most countries is under severe and increasing pressure from rapid urbanisation, pollution, tourism development, over exploitation of coastal resources (specially biological resources) and continued development in hazard prone areas. Repeated and consistent conflicts exist particularly in resource allocation and approach towards utilisation and exploitation of these resources. The coast provides important locational benefits for heavy industries, energy generating facilities, fuel and raw material processing etc. Coastal landforms have an important role in protecting the coastline from erosion and flooding. A large portion of wave energy arriving at the coast is absorbed by the salt marshes and mangrove swamps, if present. Besides, Spit and bar perform similar role in dissipating wave energy. Coastal wetlands are home to a large variety of birds, plants and other biota and also serve the important role of filtering impurities in the water coursing through them.

Despite its fragility, the coastal zone is amazingly resilient. The ecosystem as a whole is a dynamic and regenerative force, if left alone, natural mechanism operate to maintain an equilibrium between all living and the natural environment (Beatly et al., 1994). There are limits, however, to the extent of coastal ecosystem can withstand external assaults to its integrity. That is why, proper assessment of coastal resources as well as dynamics working in this region need to be studied in detail. This is of utmost importance for any nation considering the huge resource potentiality vis-à-vis the indiscriminate way of exploitation of this region.

Background of the study

The definition of the coastal zone depends upon the purpose at hand. From both the management and scientific view points, the extent of coastal zone will vary according to the nature of the problem, the extent of the resource and the administrative boundaries as imposed by the Government with jurisdiction and responsibility for management in the coastal zone, a very dynamic as well as fragile system. This is a result of interactions with the atmospheric processes, the operations of different ecosystems and coastal processes, and upcatchment activities as well as the different degrees of development activities being carried out. The interaction between the coastal zone and the high seas is a further aspect of this dynamism, particularly in relation to living marine resources and the impact of environmental degradation and pollution. The boundaries of coastal zone should extend as far inland and also towards sea as necessary to achieve the objectives of the management programs. Present sustainability of these resources should be measured before policy formulation. Detail knowledge of the present land use/ land cover pattern, both micro and macro level resource survey and landform-natural processes interaction study is needed before policy formulation for sustainable development. Policy development and implementation must take into account the dynamic nature of these systems and the specific local and regional character of coastal resources and processes. Detail plan can be prepared only on a case by case basis for specific regions.

COASTAL REGULATION ZONE OF INDIA

The Coastal Regulation Zone (CRZ) notification of 19th February, 1991 is the direct outcome of Environmental Protection Act, 1986. Notification under section 3(1) and 3(2)(v) of Environmental Protection Act, 1986 rule 5(3)(d) of the Environment (Protection) Rules, 1986 states the specific portion of coastal stretches as Coastal Regulation Zone and also mentioned about the Regulating Activities in the Coastal Regulation Zone. According to the notification, the coastal stretches of seas, bays, estuaries, creeks, rivers and back waters which are influenced by the tidal action (in the landward side) up to 500 m from the High Tidal Line (HTL) and the land between the HTL and LTL (Low Tide Line) as Coastal Regulation Zone and imposes, with effect from the date of this notification, the restrictions on the setting up and expansion of industries, operation or processes etc. in the said Coastal Regulation Zone (CRZ). For the purpose of this notification, the High Tide Line (HTL) will be defined as the line upto which the highest of high tide reaches at spring tide. It has also been mentioned that the proposed regulations will apply in the case of rivers, creeks and backwaters, may be modified on a case by case basis for reasons to be recorded while preparing the Coastal Zone Management Plan (CZMP) and, however, this distance shall not be less than 100 meters or the width of the creek, river or backwater whichever is less.

This notification has been amended in the subsequent period for a number of times. However, the basic premises remain the same and it has been decided that the distance from the high tide line (HTL) shall apply to both sides in the case of rivers, creeks and back waters and may be modified on a case by case basis for reasons to be recorded while preparing the coastal zone management plans (CZMP). However this distance shall not be less than 50 meters or the width of the creek, river or back water whichever is less. Further, the distance up to which development along rivers, creeks and back waters is to be regulated shall be governed by the distance up to which the tidal effect of sea is experienced in rivers, creeks or back waters, as the case may be, and should be clearly identified in the Coastal Zone Management Plans. Mapping of HTL is to be done uniformly over all the coastal states so as to avoid conflicts in management treatment under the same regulation in different areas. Surveyor general of India has been vested with the responsibility of demarcating HTL by Ministry of Environment & Forests and the work is in progress. Mean while, even before finalisation of the High Tide Line, an attempt has been made to create a baseline data regarding the mode of utilisation of the area falling within the Coastal Regulation Zone of West Bengal with the help of published maps of Survey of India and employing the Remote Sensing technique.

General Geology and Geomorphology

The study area i.e. Coastal Regulation Zone of West Bengal is the eastern and southern part of Bengal Geosyncline (stable shelf in the west and deep basin in the east) bordered in the west by the Indian shield which is separated by a series of burried basin marginal enechelon faults. Intensive Geophysical surveys and deep drilling data in this alluvial plain of West Bengal revealed that the extensive subaquous basaltic lava (as Rajmahal Trap) is overlying Permo-Carboniferous coal bearing Gondwana sediments of continental environment. Later geological history is the repeatation of marine transgressions and regressions which were the major phenomena controling the depositional environment during the evolution of this basin. The sediments overlying the trap were deposited mainly in the continental environment. The drainage system originated from Indian shield (Chhotanagpur plateau part) were the then geological agents of sediment transportation and deposition. The first evidence of transgression in the basin is during Cretaceous but upper Cretaceous was a phase of regrssion. Other phases of transgressions are early Palaeocene (local), Eocene (extensive), late Oligocene-Miocene (extensive), and Pliocene(two phases - early and late Pliocene). Late Palaeocene, late Eocene early

Oligocene, mid Miocene and upper most Miocene were the phases of regressions. Although the Quaternary was in general in regressive phase but oscillatory environment was also there at that time (Tiwary & Banerjee, 1985). The Eocene Hinge Zone, a zone of tectonic flexure and fault, passing Calcutta-Ranaghat-Mymensingh separate the relatively stable shelf from the deeper basin of the geosyncline. Post Eocene eastward tilting cause very thick pile of stratigraphic horizons in the eastern part.

A detailed study of landforms and soils by air photo interpretation and extensive field work was carried out by Niyogi (1970 and 1975). He identified four successive formations namely, Laterite upland, Older deltaic plain, Younger deltaic plain and Recent deltaic plain. According to him, delta building by existing set of rivers continued through the Quaternary there by pushing the shore line steadily eastward or southeastward. He also advocated that due to continuous uplift of the region in the west, each younger formation was deposited further seaward than the preceding one thereby leaving a portion of the latter uncovered and at each junction, the older formation dips below the younger, the regional dip being towards east or south-east. He also suggested that in the present area, a steady but slow uplift continued all through the Quaternary; due to this phenomenon, belts of land remained uncovered to the west during successive high stands. It is also apparent, according to Niyogi, that the cumulative effect of the uplift operating for quite a long time progressively steepened the surfaces of older landforms all of which were initially much flatter. Deltaic plain of modern delta of comparable lithology slopes at 0.07 m per km. while Ganga delta slopes 0.1 m per km. towards south (Mallick, 1971).

According to Chakraborty (1991), the successive rows of dunes (viz. Ancient, older and Beach front dunes) with intervening clayey tidal flats and local sandy beach ridges are indicative of punctuations in the regression of the Holocene sea in this area. He also suggested that the "Ancient dune complex" all along the Medinipur coastal plain (about 10 to 15 kms. north of the present day shore line) indicates the position of the ancient strand line in the area. The C^{14} dating of the sediments from ancient fluvio-tidal flat (5760 ±140 YBP) bordering the Ancient dune complex in the south confirms that the higher strand line in the post glacial (Holocene) period is at present represented by the "Ancient dune complex" which is around 6000 YBP - the optimum Flandrian transgression. The C^{14} dating of sediments from ancient inter tidal flat (just south of Ancient dune complex) gives an age of 2920 ± 160 YBP which indicates the first punctuation in the regression of Holocene sea in the area under consideration.

Subsequently, Chakraborty (1995) further reported the presence of a morpho-structural lineament in the same alignment and as an extension of the continuous dune ridge (the Ancient dune complex) in the eastern part of the Hooghly river. He has identified this lineament on the basis of study of the Landsat (both black & white and FCC/TM) and reported that the said lineament passes in a north easterly direction and separates the lower deltaic plain of the Ganga-Brahmaputra in two sectors. According to him, the area to the south of this line are characterised by the presence of islands (draped with mangroves) separated by an "embroidery" network of active tidal creeks and islands. He concluded (from C^{14} dating of the sticky grey clay sample of inter-distributory mangrove marsh unit of Namkhana area which gives an age of 3170 ± 70 YBP) that the lineament in the eastern part of Hooghly river estuary is nothing but a mark of ancient strand line. During our present study, although such a zone of separation of tidal creeks have been noticed, but specific identification of such a lineament in the imageries used has not met any success.

Climate

Tropical humid climate with annual rainfall of about 1650-1800 mm in central and northern areas, and as much as 2790 mm on the outer coast is observed along this coast. The mean maximum temperature is 29^0C (December-January). Average humidity is about 80% (Chaudhuri & Chaudhury, 1994). South-West Monsoon enters in mid June and continues up to October. During winter months, when North -East Monsoon prevail (October to mid March), lowest precipitation is noticed.

Cyclonic hazards are frequent which cause destructive surges along the coast. On an average, three to four cyclonic storms form in the Bay of Bengal mostly during pre monsoon (April-June) and post monsoon (September - November). These spirally moving cyclones or typhoons developing in Bay of Bengal strike Bangladesh, West Bengal, Orissa, Andhra Pradesh and Tamilnadu coast with average wind velocity of 100-120 km/hour and sometimes cause heavy down pour in the order of 15-25 cm within the duration of passage of cyclones over the belt. A study carried out by Saha et al. (1984) indicates that over a period of about 100 years (From 1889 to 1970), frequency of cyclonic surge, as noted along the East Coast, is maximum at the coast of 24-Parganas district of West Bengal. Incidentally, mangrove forest is present all along the coast of 24-Parganas district providing protection to the coast from severe surge waves.

The storms give rise to spectacular dune landforms on Balasore-Contai coast. Extremely irregular and high waves are generated and severe beach erosion took place, particularly if the storms happen to coincide with either new moon or full moon high tides. Immediately after such cyclones, the beach is left in a broken state having deep channels at an angle to the beach slope. Within few days these irregularities are smoothed out. Surges bring coarser sands from the offshore and removed finer sands to the deep waters. Thus, ultimately the beach gradually gets lowered.

The estuarine zone experiences a relatively weak tide during winter months. Strong tides prevail during February to May followed by a short period of relatively weaker tides and returning to strong tides during pre-winter season. During summers, strong tides govern the channel regime and influence the upstream sediment movements. But during monsoon rains, the channel regime is determined by the interaction of head.water discharge and the tides, which together influence the seaward drift of the sediments. Sedimentation during this period cause congestion of the channel in its mouth region. Salt concentration in this estuarine environment fluctuate greatly and the density difference between saline water and fresh water at their interface creates current which plays a major part in the formation of estuarine landforms. Again, the flood tide is short in duration, lasting two to three hours, while the ebb flows for the remaining eight or nine hours of the twelve hour tidal cycle. One effect of this disparity is that flood tide velocities are much greater than ebb tide and each flow can take a different pathway through the maze of minor channels within the estuary (Chaudhuri and Chaudhury, 1994).

Study Area

The 220 km. long coastal stretch (between $87^0 26'$ E to $89^0 08'$ E) of West Bengal extends from Subarnarekha estuary on west to Harinbhanga river in the east. The Coastal Regulation Zone of West Bengal is the Lower Ganga Delta plain. This stretch of coastal Quaternaries exhibits varied geomorphological signatures (sand dunes, beach ridges, flood plains, tidal shoals etc.) evolved out of dynamic and varied interactions of river and marine agencies.

Classification System

For detail up-datation of present landuse/landcover pattern with special emphasis on built-up areas, tidal wetlands, industrial activities and sand/rock mining areas the following classification system (Table 19.1) was adopted for the present work. In arriving at the classification and nomenclature the following criteria were considered.

1. This classification system is applicable over large areas under study i.e. through out Indian coast.
2. The minimum interpretation accuracy and reliability of different landuse/ landcover classes from satellite data is greater than 85 %.
3. The nomenclature, definition and framework to the extent possible, should be compatible with existing terminology adopted by different departments.

METHODOLOGY

The present work attempts to prepare land use/land cover maps, with special emphasis on built-up area, tidal wetlands and sand/rock mining areas in CRZ of West Bengal on 1:25,000 scale. Considering the efficacy and cost effectiveness of Remote Sensing, it was decided to utilise this technology.

Visual interpretation of SPOT I HRV 1 & 2 MLA multi-spectral FCC (bands 123) data of 1988-1989 on transparent sheets on 1:4,00,000 scale using PROCOM II has been accepted as the mapping technique for the present work. IRS 1B LISS II FCC (bands 234) data of 1992-93 on transparent sheets on 1:5,00,000 scale have also been consulted during interpretation for the areas for which SPOT data was not available.

Table 19.1: Classification System for Coastal Land Use Mapping

Category	Level of Classes
Agricultural land	1
Forest (Non-tidal)	2
Natural	2.1
Man-made	2.2
Wetland	3
Estuary	3.1
Lagoon	3.2
Creek	3.3
Backwater/Kayals	3.4
Bay	3.5
Tidal flat/Mud flat	3.6
Sand/Beach/Spit/Bar	3.7
Coral reef	3.8
Rocky coast	3.9
Mangrove	3.10
Dense	3.10.1
Sparse	3.10.2
Salt-marsh / Marsh vegetation	3.11
Other vegetation (scrub/grass/algae/seaweeds)	3.12
Barrenland	4
Sandy area/dunes	4.1
Mining areas/dumps	4.2
Others (rock outcrops/gullied eroded/badland)	4.3
Built-up land	5
Habitation	5.1
Habitation with Vegetation	5.2
Open/Vacant land	5.3
Transportation	5.4
Roads	5.4.1
Railways	5.4.2
Harbour/Jetties	5.4.3
Airport	5.4.4
Waterways	5.4.5
Other features	6
Reclaimed Area	6.1
Salt pans	6.2
Aquaculture ponds	6.3
Pond/Lakes	6.4
River/Streams	6.5
Drains/Outfalls/ Affluents	6.6
Seawall	6.7
Embankment	6.8
High waterline	6.9
Low waterline	6.10

The data used are mostly acquired during low tide condition since during low tide condition, most of the coastal wetlands are exposed and thereby easy to identify. Further, since the area under consideration, is covered by clouds during monsoon periods, data acquired during the period October to March have been utilised. However, in certain cases even the data of other periods have also been consulted.

Firstly, the topographic sheets, published by Survey of India on 1:50,000 scale or 1:63,360 scale, was photographically enlarged to 1:25,000 scale in parts and then collated to prepare an integrated map of the same area on the desired scale. Through this process, the likelihood of optical distortion had been minimised and distributed throughout the map. This method has been found to be very convenient for achieving planimetric accuracy. The scale converted map was then used for preparation of base maps on which the preliminary interpretation of Remote Sensing data were carried out using PROCOM II optical enlarger.

Various prominent and permanent features like roads, man-made canals, embankments, even creeks (particularly in case of mangrove region) etc, identifiable in images have been drawn on the base maps in order to make correlation between images and ground. Remote Sensing data were all obtained from the National Documentation Center, National Remote Sensing Agency, Hyderabad. An image interpretation key has been developed with the help of the extensive ground truth information and the Survey of India topographical maps on 1:50,000 scale.

Preliminary interpretation of RS data on transparencies on 1:4,00,000 scale were enlarged sixteen times using an optical enlarger known as PROCOM II while the IRS 1B LISS II transparencies on 1:5,00,000 were enlarged twenty times using the same enlarger. The individual categories were delineated and classified according to the classification system already mentioned. Categories of size less than 2 mm X 2 mm have not been mapped. All the doubtful areas were identified and listed for ground truth verification. Detail ground truth survey was carried out in these areas. The field information was used in modifying the subclasses and drawing correct boundaries. Through this exercise, the image interpretation key has also been further developed. With the help of this image interpretation key, final interpretations were carried out.

Landuse maps of Coastal Regulation Zone, thus prepared were made ready for cartographic work. Area calculations of the individual units were then undertaken, using a digital planimeter (PLACOM Kp-90N of Koizumi, Japan make). The average of three consecutive readings was recorded as the physical area of that particular unit.

RESULTS AND DISCUSSIONS

Based on tidal amplitude, the West Bengal coast can be sub-divided into two different coastal environment, namely :-

i) the macro tidal (tidal range : > 4 m) Hooghly estuary, characterised by an embroidery network of tidal creeks, encompassing the islands and offshore linear tidal shoals, aligned perpendicular to the shore line and separated by swales.

ii) the meso tidal (tidal range : 2-4 m) Medinipur (Digha - Junput) coastal plain (a part of Balasore - Contai coastal plain) to the west of the Hooghly estuary, with successive rows of dunes with intervening clayey tidal flats.

Two unique coastal ecosystems, namely deltaic mangrove-swamp coast of the eastern part and dune-interdune coast of the west, coinciding with two above mentioned different tidal environment, is separated by the Hooghly estuary. The landward limit of coastal zone i.e. the width of this zone has to be ascertained before the formulation of any management plan.

For the present study, CRZ boundary, in all cases, has been considered to be 500 meters away from HTL irrespective of the width of the river/creek/canal/back water bodies. Theoretically, Coastal Regulation Zone (CRZ) extends landward upto which the coastal processes are active. However, for all practical purposes, in the present study, CRZ has been drawn along the both sides of the river, canals, creeks etc. to the extent within inland upto which tidal symbols have been marked on Survey of India topographic sheets, although in few cases, tidal influence have not been noticed in such streams/ canals/ creeks etc. during the field work.

In-spite of the negative field observations, these areas have been included since, (i) ground truth has been collected on a particular day rather than through out the year; (ii) the CRZ has been drawn strictly following Survey of India maps where tidal symbols have been demarcated; (iii) the experience of local people regarding the tide conditions have also been considered. Again, it has been observed from images that in certain areas, the choking of the river bed is very prominent and there is hardly any possibility of tide to play. These areas have been excluded from CRZ.

Further, certain areas have been included under CRZ of West Bengal along certain canals, streams or creeks although in Survey of India topographic sheets, tidal symbols have not been marked in those particular stretches. These area were included specifically on the basis of ground truth survey and image signature.

On the basis of landuse / landcover unit characteristics and nature of the tidal influence the entire Coastal Regulation Zone of West Bengal may be subdivided into following sectors :

i) **Deep-inland sector:** It represents the innermost CRZ in the landward side mainly along the rivers Hooghly and Ichhamati beyond Tribeni and extends to the farthest point upto which tidal symbols have been marked on Survey of India topographic sheets. As per report of Calcutta Port Trust, tidal activity is still present upto Tribeni although in Survey of India topographic sheets, tidal symbol has been marked beyond this point.

ii) **Central sector:** It represents the area in between Diamond Harbour and Tribeni. As per National Institute of Oceanography (1987, the proceedings of the 9th International Symposium on Tropical Ecology), penetration of salt front along the river Hooghly never extends beyond Diamond Harbour demarcating the estuarine limit, although tidal activity persists upto Tribeni.

iii) **Sunderbans sector:** It represents the area in Sunderban region from the open sea to Dampier-Hodges line. The famous Sunderban mangrove region falls within this sector.

iv) **Medinipur coastal sector:** It represents coastal part of Medinipur district on the western side of Hooghly estuary.

Different landuse/landcover units, as identified within the CRZ of West Bengal, have been delineated and mapped on 1:25,000 scale. Details of these units are discussed below.

Agricultural Land

This comprises areas primarily used for cultivation such as rice, fiber (jute) crops, fruits etc. Further classification of this unit has not been done according to the particular type of cultivation. Again, some pockets of fruits and/or vegetable gardens have not been included in this category, since separation of such lands occurring within the habitated region was not possible. The area covered by this unit in four different sectors is given below (Table 19.2). From the table, it is clear that so far as the agricultural areas in terms of percentage of the total CRZ of the respective sectors are concerned, it is maximum in Medinipur coastal sector and least in Sunderban sector. This may be due to the fact that in Medinipur sector, during and after the rainy season, the surface drainage of rain water is comparatively better than in case of Sunderban sector, as a consequence of which depth of standing water over the paddy fields is comparatively low in Medinipur sector which has been considered as a favourable condition for paddy cultivation. Besides this, least saline water incursion, limited reclamation for industrial and urban growth are also responsible for that. While saline water incursion, water logging over huge area during monsoon, extensive presence of mangroves in Sunderban Reserve Forests in the southern part of the sector are the major reasons behind the comparatively least percentage of land within CRZ among the four sectors being used for agriculture

Table 19.2: Different Coastal Zone Sectors

Sector	Area in km^2	Area in %
Deep-inland sector	157.766	55.92
Central sector	620.996	53.92
Medinipur coastal sector	378.832	62.01
Sunderban sector	1507.643	31.12
Total Agricultural Land	2665.237	38.69

Note: Percentage of this particular land use type in different sectors have been calculated on the basis of total CRZ-area of the respective sector.

The agricultural field, in Sunderban sector, is lying below the river or creek bed level due to immature reclamation of the tidal flood plain by constructing embankments along the creeks. It should be mentioned in this respect that the agriculture field of Deep inland sector and Central sector are mostly jute growing fields but in other two sectors it is used for extensive paddy culture. Vegetables are mainly grown in Sunderban and Central sectors whereas cashew nut is mainly grown on sandy area of Medinipur coast.

Forest

Both tidal and non-tidal forests are found in Coastal Regulation Zone. Tidal forest is the predominant unit and this has been included under the Wetland unit. About 3.595 and 12.426 km^2 area is covered by the natural non-tidal and man-made non tidal forest respectively. Cassurina, Akashmoni, Sisoo, Babla etc. are the main species of planted non tidal type. At places, plantation of mangroves are also found viz. Bogkhali, Agnimari Char etc. The Sunderban Reserved forest has been considered under the mangrove units.

Wetland

The wetlands of coastal regulation zone in West Bengal comprises estuaries, creeks, backwater, tidal flats/mud flats, sand/beach/spit/bar, mangrove forest(dense and sparse), salt marsh and other vegetation. It covers about 36% (Total area = 2467.146 km2) of the total CRZ area of West Bengal. Distribution pattern and other details of different types of wetlands within CRZ of West Bengal have been discussed below.

Estuary

The river Hooghly forms a big estuary near its mouth which is the major water way for ships of Calcutta port and Haldia port, boats and launch. Other major estuaries found to the east of Hooghly estuary are the Saptamukhi, the Thakuran, the Matla, the Bidya, the Gosaba, and the Harinbhanga. Besides these numerous smaller estuaries occur in this sector. In Medinipur part, mouth of Rasulpur forms the only estuary. These estuaries are highly productive areas of living marine resources.

Creek

Dense embroidery network of creeks fed Sunderban area which are influenced by daily tides. The high erosive energy of storm surges is absorbed by the network of creeks. The creeks are gradually filled with sediments and getting choked from upper reach.

In ancient times, all the major rivers of Sunderbans used to receive remarkable quantity of fresh water from Bhagirathi - Hooghly-Padma river. But, be-headation of these channels caused by greater rate of subsidence of the deeper geo-synclinal eastern part as well as easterly tilting by neo-tectonic movement during the 12th century (Morgan & McIntire, 1959) made this channel system as tidal drainage with very insignificant freshwater discharge during monsoon period. Again, reclamation of this immature flood plain by constructing embankments almost all along the tidal channel lead to imbalance growth where river beds stand higher than the reclaimed land which is being used for agriculture.

Construction of embankments on both sides of the major rivers/ creeks bring forth dead channels by degrading smaller creeks on both sides. These degraded channels with time have been transformed into significant wetlands in Sunderbans and play an important role in the drainage of the area. The distribution pattern of different units of wetland in four sectors is given in Table 19.3.

Table 19.3: Distribution Pattern of Different Units of Wetland

Units	Area	Deep inland Sector	Central Sector	Sunderbans Sector	Medinipur Sector
3.6	In km2	0.475	28.183	181.994	13.295
	% of total crz	0.17	2.45	3.76	2.18
3.7	In km2	0.426	0.057	19.602	11.824
	% of total crz	0.15	Negligible	0.40	1.94
3.10.1	In km2	-	-	1915.538	-
	% of total crz	-	-	39.54	-
3.10.2	In km2	-	-	233.458	0.506
	% of total crz	-	-	4.82	0.08
3.11	In km2	0.644	-	21.896	-
	% of total crz	0.23	-	0.45	-
3.12	In km2	2.081	9.155	24.362	3.656
	% of total crz	0.74	0.79	0.50	0.60
Total	In km2	3.626	37.395	2396.844	29.281
	% of total crz	1.29	3.25	49.48	4.79

Note: Percentage of this particular land use type in different sectors have been calculated on the basis of the total CRZ-area of the respective sector.

Tidal flats/Mud flats

Mud flats/Tidal flats is one of the most important category of the landuse / landcover feature in Coastal Regulation Zone (CRZ). In the western part of West Bengal coast i.e in Medinipur coastal sector it is sandy but in Sunderban sector and in other two sectors it is muddy in nature. Three types of mud flats are found namely ; Saline water mud flat, Brakish water mud flat and Fresh water mud flat. In two sea facing sectors i.e. in the Sunderbans and Medinipur coastal sectors, the mud is very saline and major portion is inundated by daily tides while mud flats of central and deep-inland sectors get inundated by pushed back freshwater during spring tide and monsoon period only. Out of 223.947 km2 area (excluding Mangrove swamp area) about 81.27 % (181.994 km2) of the total mudflat/tidal flat area within the CRZ is restricted in Sunderban sector and Medinipur coastal sector constitute about 5.94 % (13.295 sq. km.) of the total tidal flat/ mud flat. The rest lies in central sector (28.183 km2) and deep-inland sector (0.475 sq.km). Mud flat of central sector is restricted to the recent chars of Rupnarayan river, Kaliaghai river, Kapaleswari river and Haldi river.

In Sunderban sector mud flats occur as elongated strip on both sides of the tidal river and creeks, as patches within mangrove swamp and along the periphery of the islands. A remarkable part of this highly productive unit is now under aquaculture bheries. The vast mud flat of Canning area has been reclaimed for traditional practice of brakish water aquaculture, particularly for culture of *Paneous Monodon*. At places, mud flat area has been converted into vegetable garden. In certain areas linear patches of sand are seen lying within the mudflats as has been noticed in Jamudwip. Even the wetlands, particularly mudflats within the CRZ are not spared. These areas are consistently under threat of reclamation. As for example, a vast stretch of mudflat (6.644 km2) lying in between Kaliaghai river and Kapaleswari N near Daspur, Medinipur is normally under water for about six months during monsoon and post-monsoon period. This wetland is being destroyed by pumping out the stored water in post-monsoon period for the purpose of Boro-cultivation. The mudflats are also being used for brick industry, specially in central and deep inland sector where the quality of mud is suitable for preparation of bricks.

Sand/Beach/Spit/Bar

Beaches in West Bengal coast are sandy or muddy. Sandy beaches are noticed in Medinipur district on the western side of Hooghly river whereas vast extension of muddy beach is found in 24-Parganas part on the eastern side of Hooghly, specially to the east of Bogkhali. Well developed beach is found near Digha, Dattapur, Shaympur, Dadanpatra, Baguranjalpai, Dariapur, Nij Kasba, Bamanchack, Beguakhali, Gangasagar, Dublat, Chemagari etc. In other parts, beaches are narrow and it is almost absent on South-Eastern part of the coast, since it is covered under mangroves. Exposition of older mudflat in Bogkhali coast and Sankarpur coast is the clear indication of beach erosion. Formation of spit is found near Junput at the mouth of Pichaboni khal. Another minor hook shaped projection is found at the mouth of Sankarpur harbour. The beach front dune complex of Medinipur coast has been included under this unit. Only 0.40 % of the total CRZ area of Sunderban sector is covered by sand and these are mostly narrow linear ridges occuring on the narrow muddy beach or at the peripheral part of the island viz. Jambudwip. Submerged bars and chars consisting of sand, silt and clay are found all along the coast and rivers namely – the Hooghly, the Matla, the Thakuran, the Saptamukhi, the Gosaba, the Raimangal, the Harinbhanga etc. Few of these partly covered with marsh vegetation as well as mangrove swamp at places. These shoals are very unstable changing frequently their position and shapes specially during SW monsoon.

Mangrove

Mangrove swamp is the major unit of vegetated wetland in CRZ of West Bengal. The total area covered by such wetlands is about 2149.496 km2 (31.20 % of the total CRZ area). From the satellite image of 1992 dense mangrove area and sparse mangrove area has been calculated as 2098.00 km2 and 129.19 km2 respectively. The major part of this mangrove area is now under Sunderban Biosphere Reserve. This mangrove colony is the largest single block of tidal halophytic mangroves of the World. They are in the form of dense forest along east and south-east parts of the coast while small patches of mangroves are found in Sagar island, Gangadharpur, Nijkasba and man-made mangrove units are also found in few pockets viz. Agnimarichar, Bogkhali, Sagar island etc. Recently mangrove steps in few newly formed island viz. Balaribar island.

Mangroves of Sunderbans are the most vulnerable units amongst all other units of CRZ in West Bengal. In spite of its immense ecological importance, the clearing of mangroves is going on, in certain pockets, particularly for shrimp culture. As for example, about 10.59 km2 of mangroves has been cleared around Charkhali area, South 24-Parganas district. Such types of clearing can also be noticed in Chuprijhara, Vijoynagar, etc. of South 24-Parganas. But, in certain areas, it has also been observed that disappearance of mangroves is due to natural causes, may be accentuated by anthropogenic interventions.

Salt Marsh / Marsh Vegetation

This landcover unit cover about 22.54 km2 area of which about 21.896 km2. is restricted to the Sunderban sector. Remaining 0.644 km2 area is present in Deep-inland sector which is fresh-water marsh and occur at the mouth of recently formed ox-bows of Hooghly River near Purbasthali and Fakirdanga. Industrial and agricultural pressure on these highly fertile recent alluvial field have reduced the total area of fresh-water marsh in Deep-Inland and Central sectors. Cultivation of Jute, rice (during dry season) etc. by cleaning these marshy wetland vegetation is very common practice. Salt marshes occur in Agnimari Char, Kankramari Char, Lohachara Char, Suparbhanga Island, Amtali etc. Again, appearence of salt-marshes have been noticed in the images of IRS IB 1992 data near Khejuri and at Rasulpur river mouth. But, it should be mentioned here that the tropical marshy vegetation of these newly appearing shoals/islands normally replaces by mangrove species within few years.

Near the mouth of the Hooghly and the north-west of Sagar Island, an human habitated island, Lohachara, has been completely eroded and marsh vegetation has been appeared on this newly converted intertidal mud flat. *Cryptocornye ciliata*, *Salicornia brachiata*, *Acanthus ilicifolius*, *Swaeda maritima*, *Sesuvium portulacustrum*, *Aeluropus lagopoides*, *Porteresia coarctata* etc. are the marsh species found in these salt marshes. Studies on succession of plants on mud flats by Mukherjee and Mukherjee (1978) in eastern, central and western sector of the mangrove area revealed that *Porteresia coarctta* was the pioneer species; whereas *Acanthus ilicifolius* and *Porteresia coarctata* in the central sector and *Suaeda maritima*, *Salicornia brachiata* were the dominant pioneer species in the western sector. The fresh water marshes are *Apponogeton appendiculatus*, *Trapa natans*, *Eichhornia crassipes* etc.

Other Vegetation

The vegetated areas of unidentified species have been included under this category. The species nature is widely varying place to place. The total covering area by this category is about 39.254 km2.

Sandy Area/Dune

Three parallel spectacular dune complexes, namely Contai Dune Complex / Ancient Dune Complex, Ramnagar Dune Complex and Beach Front Dune Complex, are major landcover units of coastal Bengal. These rows of dunes are stretching from Subarnarekha estuary to Hooghly estuary. Among these three, the northern most dune complex, Contai Dune Complex, is well developed and it is about 6000 (C^{14} dating) years old. Two important cash crops, cashewnut and coconut, are grown on this sand cover area. Altogether, more than 90 km2 area is covered by the dunes but only 5.84 km2 of these dune complexes fall within the CRZ. This is simply due to the fact that the major part of this land cover unit is lying out side the CRZ i.e. beyond the line of 500 metre from High Tide Line (HTL). Even some of the area of these dune complexes falling under CRZ have been identified according to the specific landuse pattern being practiced over these complexes such as Habitation, Sand/Beach, Agriculture, Man-made forest etc.

Built-up Land

High productivity, even topography and equable climate have attracted human beings from the very begining of the growth of this Lower Ganga Deltaic plain to migrate eastward and southward. Industrial development, introduced by the British, cause rapid increment in population density during 19th century. Again, post independent industrial expansion during the five years plans and migration of people from neighbouring Bangladesh bring forth rapid cxpansion of habitated area reclaiming immature flood plains of Hooghly-Matla-Harinbhanga drainage basin.

In Coastal Regulation Zone (CRZ) of West Bengal about 1301.603 km^2 area is covered by the unit of builtup land and it is about 18.89% of the total CRZ area. The distribution of different subunits in four sectors is given below in Table 19.4.

From the mapping work it is clear that maximum habitated area is located in Sunderban sector(545.191 km2) but in terms of percentage of the respective area it is only 11.26 % whereas in Deep-inland sector and Central sector it is 40.88 % and 41.43 % respectively. Growth of industry (Hooghly industrial belt), trade and Commercial centres around Calcutta may be the prime factor behind this uneven distribution. In the Central sector, the habitated zone on both eastern and western natural levee of the river Hooghly, from Gardenreach to Tribeni, is characterised by dense population with insignificant presence of vegetation. On the other hand, in Sunderban sector, sparse linear settlement is found to exist on both sides of tide-fed river/creeks and roads. The waterways of these tidal creeks are the major transport system in this highly inaccessible area. Some vegetable growing fields, specially in Sunderban sector have been included under the unit of "Habitation with vegetation" since demarcations of such small vegetable gardens are hardly possible.

Table 19.4: Distribution of Different Subunits in Four Coastal Sectors

Units	Area	Deep inland Sector	Central Sector	Sunderbans Sector	Medinipur Sector
5.1	In km2	0.926	99.533	7.031	-
	% of total crz	0.33	8.64	0.15	-
5.2	In km2	114.313	375.90	537.11	162.186
	% of total crz	40.52	32.64	11.09	26.55
5.3	In km2	-	1.776	1.05	0.257
	% of total crz	-	0.15	0.02	0.04
Total	In km2	115.239	477.21	545.191	162.443
	% of total crz	40.85	41.44	11.25	26.59

Note: Percentage of this particular land use type in different sectors have been calculated on the basis of total CRZ-area of the respective sector.

Other Features

Reclamation of mangrove areas is very prominent in Sunderban sector. During British rule, systematic leases were granted to settlers to settle in the area, particularly from 1831 to 1833, to the south of Dampier-Hodges line. This colonisation policy encouraged the reclamation of the then mangrove area from Haroa, Minakhan in the north-east of Sagar, Kakdwip and Patharpratima on the south-west (Pargiter,1934). The low lying immature flood plain of cleared mangrove had been protected from tidal ingress by construction of embankments. This reclamation gave rise to a number of environmental problems.

Table 19.5: Land Use Change in Sunderbans of West Bengal

Landuse Type	1880	1900	1920	1940	1960	1980
Total arable & Settled Land	14570	16140	15830	16710	20820	21860
Wet Rice	11240	11560	12620	12120	15490	16040
Forest Woodlands	680	630	640	600	440	400
Grass-shrub	800	500	820	1500	1430	1790
Barren Land	500	500	500	500	500	500
Total Wetland	15160	13960	13940	12410	8530	7180
Mangrove Tidal	6149	5574	5402	5301	4231	3909
Surface Water	5740	5740	5740	5740	5740	5740

Source: Mangrove of the Sunderbans, vol. one : India; IUCN Publications

In 1829-30, Hodges and Dampier surveyed and defined the boundary of the Sunderban forest, from the Hooghly to the eastern limit of the forest. From 1830 to the early 1900's about 3737 km2 of the total forest area of 7908 km2, located to the south of Dampier-Hodges line, had been cleared for cultivation and Settlement (District report of 24- Parganas,1951). Another report made in 1872 showed that the total area under cultivation was 2783 km2, of which two-thirds was reclaimed between 1830 and 1872. By the early 20th century, over 5000 sq. km. had been brought under cultivation. The land use change in Indian Sunderbans in 1880-1980 is given below as described by Williams, 1991; which shows the mangrove depletion during this time.

However, in our present work, very recent reclamation (after 1968-69) in CRZ area has only been considered. In Sunderban sector the land has been reclaimed for aquaculture (specially shrimp culture), settlement and agriculture; where as, in Medinipur coastal sector it is for shrimp culture and salt panning. Again, in central sector, the land has been reclaimed for industrial development and related agriculture. Freshwater aquaculture is mainly practiced in deep-inland sector.

But, present study also shows that the rate of mangrove depletion in Sunderban sector has been checked considerably after taking various conservation and management programs. Since degradation of mangrove in a prograding delta is a natural phenomena so it is very difficult to check the degradation completely. As for example, remarkable cut-off of upland freshwater discharge though different river channels of Sunderban is one of the causes of depletion of few mangrove species population in this sector.

A comparative assessment from the present day imageries with the Survey of India topographic sheets points to the fact that about 51.69 km2 of new mangrove colony has been found to appear on several shoals/chars/along the creeks in Sunderban sector. Such areas are Kankramari Char, Agnimari Char, Lohachara Char, Chuksardwip, Char to the south of Koyabati etc.

Salt Pan

These are reclaimed inter tidal marsh land. This is a major land use unit near Dadanpatra of Medinipur coast. This salt pan is now used also for aquaculture. Presence of salt marsh vegetation species viz. *Salicornia brachiata*, *Suaeda maritima*, *Sesuvium portulacastrum*, *Acanthus ilicifolius* are frequent.

Aquaculture Pond

The fishery industries, which play a significant role in socio-economic as well as in physical environment, of the CRZ of West Bengal, are of diverse type, comprising saline water coastal fisheries, brackish water fisheries and fresh water aquaculture. Large scale conversion of mangrove forest, salt marshes, agricultural field, mudflats and fresh water lakes / beels to mono-culture fisheries have caused serious and irrevocable environmental damages. Surveys conducted in the Sunderbans indicate a total annual catch of fish of 25,000 metric tons and on an average 4,000 individuals are engaged in daily fishing (1.5 kg fish/day/fisherman). Besides this, about 40,000 shrimp seed collectors annually harvest about 540 million seed of *Paneous monodon* destroying 10.26 billion seeds of other fish and shrimp in this process. Similarly, reclamation of freshwater beels, specially in central and Deep-inland sector, for culture of selective species severely destroying the diversity of aquatic species as well as the whole. Another problem is that the soil become extremely acidic due to exposition of pyrite, present in the soil of mangrove area, to air and, thus, release sulphuric acid which reduce the fertility of soil greatly.

About 356.09 km2(5.17%) area of CRZ is used for aquaculture of which 352.49 km2 area is of saline and brackish water. Sunderban sector constitute 347.461 km2 area of saline water and brackish water aquaculture whereas the freshwater fisheries are restricted mainly to the beels/lakes in central and deep-inland sectors. The distribution of salt pans, reclaimed areas, aquaculture ponds, ponds/lakes in four different sectors is shown below.

Table 19.6: Distribution of Different Land Uses in Four Sectors

Units	Area	Deep inland Sector	Central Sector	Sunderbans Sector	Medinipur Sector
Reclaimed	In km2	-	-	1.063	-
Area	% of total crz	-	-	10.02	-
Salt Pan	In km2	-	-	-	28.606
	% of total crz	-	-	-	4.68
Aquaculture	In km2	-	3.608	347.461	5.025
Pond	% of total crz	-	0.31	7.17	0.82
Lake /	In km2	5.199	4.252	9.298	0.493
Pond	% of total crz	1.84	0.37	0.19	0.8

CONCLUSIONS

Coastal stretches of this part of the country are extremely complex, so far as its evolution, environment and existence are concerned. West Bengal is one of the most populated states of the country. The situation is alarming, if the total CRZ of West Bengal is considered at a time. Out of a total CRZ area of 6888.814 km2, 1296.999 km2 has been occupied by either habitation or habitation with vegetation. This represents 18.83% of total area under CRZ, while agricultural activities represent 38.69% of CRZ area. Other areas under human interference namely reclaimed area, salt pans and aquaculture ponds cover 385.763 km2 area representing 5.6% of total area. Thus, at least 63.12% of total CRZ area, in some way or other, is under intense human intervention. If the other units like man made lakes, ponds etc. be considered, then this figure will further increase. In spite of this fact, even after immense losses of the coastal natural resources, many things are still left which demand immediate attention.

Considering the vast natural resources existing in this zone and the potentiality of utilisation of these resources, mechanical approach of protectionism only will not be effective here. The situation has been further complicated since several Government agencies and departments are active here. Although, after the formulation of CRZ, clearance of major projects from Department of Environment is mandatory, it may not be sufficient enough for the sustainable use of this fragile region, particularly in respect of small scale day to day activity being carried out in this zone by the local people. It would be better, if a further intensive survey can be carried out in this region before finalisation of a long term management policy for the sustainable use of these resources. Also, if a single authority can be identified to monitor the activities and the processes going on in this zone. The local people can also be involved in such activities through local bodies. Some sincere efforts are required immediately otherwise hardly anything will be left for the future.

Acknowledgement: Authors are grateful to Dr. T. S. Bandyopadhyay, Director, IWMED and Dr. R. R. Navalgund, Mission Director, RSAM and Group Director, RSAG, Space Applications Centre for their help and encouragement during the execution of this work. We are indebted to Sri P. S. Snayal, Chief Environmental Officer, Dept. of Environment, Govt. of West Bengal for his consistent support. Fund for this work has been provided by Ministry of Environment & Forests, Govt. of India through SAC and we are grateful to them.

REFERENCES

Beatley, Brower, & Schwab : An introduction to Coastal Zone Management; Island Press.

Chakraborty P(1991) : Subarnarekha delta - A Geomorphic Appraisal; Memo.Geol. Soc. of India, No.22, pp. 25-30.

Chakrabarti P(1995) : Evolutionary history of the coastal Quaternaries of the Bengal plain, India.Proc. Indian natn. Sci. Acad., 61, A, No. 5, pp.343-354.

Chakrabarti P : Process Response System Analysis in the Macrotidal estuarine and Mesotidal Coastal plain of Eastern India. Memo.Geol. Soc.India. No. 22, pp 165-181.

Chaudhuri A.B & Chaudhury A.(1994), : Mangroves of Sunderbans, Vol-1, IUCN Publications, India.

Mallick S. (1971) : Unpublished Ph. D. Thesis, Dept. of Geology and Geophysics, Indian Institute of Technology, Kharagpur.

Morgan J P & McIntire (1959) : Quaternary Geology of Bengal Basin, east Pakistan and India, Bull. Geol. Soc. of India, Vol 70, pp 319-342.

Niyogi D & Mallik S (1973) : Quaternary Laterite of West Bengal : Its geomorphology, Stratigraphy and Genesis. Geol., Min.& Met. Soc. of India, 1973.

Niyogi D (1975) : Quaternary Geology of the Coastal Plain in West Bengal and Orissa, Ind. Jour. of Earth Sc. vol 2, pp 51-61.

Niyogi D (1970) : Geological Background of Beach Erosion at Digha, West Bengal. Bull. of the Geol. Min. & Met. Soc. of India. vol. 43.

Saha A. K. et al (1984). : Report of the cyclone review committee, Department of Science & Technology, Govt. of India, New Delhi Vol.-2, 480p (Unpublished).

Tiwari S & Banerjee S.K (1985) : Seismic stratigraphy of onshore West Bengal basin; Bull. of O.N.G.C , vol 22.

20

HOOGHLY RIVER AS SOURCE OF COASTAL POLLUTION IN WEST BENGAL

A. Kumar, D.A. Patil and Rakesh Kumar

ABSTARCT

The Hooghly river in West Bengal is the major fresh water source to a host of townships and large number of industries in the Calcutta Metropolitan District. It carries the waste water from more than 300 drains within the 100 km stretch of the tidally influenced zone of the Hooghly estuary. The results of water analysis indicate that levels of trace metals especially cadmium and chromium are above the drinking water limits in some regions of the river stretch. Also, the bacteriological quality indicate the high degree of contamination of river water from wastes of domestic and animal origin. As regards to the biological indicators of water quality, the diversity index for zooplanktons varies from 0.91 to 2.78 and the total count for phytoplankton is between 300 to 1600 per 100 ml. There is no significant change in the species composition during low and high tides. The bottom sediment shows higher metal concentration in the upstream zones of the river. It is recommended that regular monitoring of Benthos in the critical zones of the estuary area be made an integral part of the future monitoring since it provides a good indication of water quality status over a period of time. Also, use of mathematical models will be of great use to facilitate optimum use of resources in the region.

INTRODUCTION

India with about 6000 km long coast line is witnessing an alarming trend of water quality deterioration in the near shore region. About 25 percent of population in India lives near the coasts (CPCB, 1993). Due to anthropogenic interference, exploitation, uses and abuses of the marine environment to meet the growth and changing needs of the modern civilization, various environmental problems have started not only in India but various countries of the world. The backlash of growth and developments, obviously fueled by the advances of science and technology, is no longer sparing the marine environment, specially in the coastal areas which were practically unaffected until recent past. Direct dumping of waste materials in the seas, waste water discharges through marine outfalls, large volumes of untreated or semi-treated wastes generated in small communities and industries ultimately find way to the seas (Salas, 1994). Intermixing of upland waters and saline waters of the sea creates a distinctive aquatic environment- a kind of buffer zone - between the terrestrial and the marine environments, which is highly variable in nature due to regular rhythmic rise and fall of the tides, the currents and the wave actions. The coastal marine environment is a complex system due to its support mechanism and its self purification capabilities (Alder, 1973).

With noticeable increase in marine pollution and the consequential marine resources, specially in the Mediterranean sea, Red sea etc. concern was expressed in the United Nations Conference of Environment in Stockholm (June 5-16, 1972) attracting global attention the urgent need of identifying the critically polluted areas of the environments, specially in coastal waters, for urgent remedial act conference unanimously resolved that the littoral states should take at the National level for assessment and control of marine pollution sources and carry out systematic monitoring of marine pollution to ascertain the efficacy of the pollution regulatory actions taken. Beginning with the resolution of the Untied Nations, Conference on Environment (1972), and in the light of the agreement of the United Nation Convention on the 'Law of the Sea' (1982), defining jurisdiction of waters, which was signed by India, a model comprehensive action been evolved under the United Nations Environment Programme (UNEP's Regional Seas Programme involves procedures for assessment problem areas and monitoring activities that are potentially causes pollution). Keeping with the international commitments and in great interest the Government of India initiated programmes for assessing of marine pollution for suitable regulatory abatement measures.

Indian Scenario

India is a country of rivers with variable discharges joining from upland and contribute to coastal pollution by transporting a wide range of pollutants through land drainage. Three out of four largest Metropolitan cities of the country, with a burgeoning population, are situated on the coast. Besides large population living in the coastal areas a large number of industries/establishments are also located in the region. The exploration and exploitation of oil from the off-shore regions such as the Bombay High and the traffic of oil tankers from Gulf countries are increasing and some times due to oil spill or leakage shores have been contaminated. The authority to control pollution of the marine coastal waters, upto 5 kms. From the coastaline into the sea, is vested with the Pollution Control Boards of the Coastal States under Section 2(I) of the Water (Prevention and Control of Pollution) Act- 1974 (GOI, 1974). There are nine Coastal States viz., Kerala, Tamil Nadu Gujarat, Mahrashtra, Goa, Karnataka, Andhra Pradesh, Orissa and West Bengal; and three Union Territories : Pondicherry, Andaman and Nicobar Islands and Lakshadip Islands.

The sources of the pollution of the coastal areas are mainly the large urban conglomerate. Some times when the waste water discharge is not directly going in the sea, it follows the route of discharge in a river and the river discharging the waste in the sea. An assessment of a situation of West Bengal has been carried out with regard to the point source discharges into River Hooghly within CMD area of West Bengal.

The coastline of West Bengal is about 200 km long and comprises of a network of flood channels. The major part of the coastline is formed by the deltaic Sundarbans supporting the world's largest mangrove ecosystem which extends up to lower Bangladesh with tiger habitat which has been declared as a "Biosphere Reserve". Ganga (Hooghly) and six other major estuaries in the region, -forming Ganga outfall system, transport the world's largest sediment load to the Bay of Bengal along with all other waste discharges brought down from urban and industrialized areas of the vast Ganga basin. This coastal stretch, with its typical mangrove ecosystem, forms a highly complex and very rich productive coastal zone. Besides the sewage of Calcutta Metropolitan and other municipal areas a very large quantity of industrial effluents find way into the coastal waters mainly through the Hooghly estuary (29 large polluting industries in the State having been identified under the Ganga Action Plan). The west of the Hoogly outfall viz. Haldia- Contai region is a highly productive and yet under- developed agricultural land which is poised for industrial development. Due to port related operations at Haldia and Calcutta port, there is a heavy traffic of oil tankers, cargo ships and mechanized fishing vessels. Digha is the major centre for tourist attraction.

HOOGHLY ESTUARY

Hooghly is the distributory of Ganga forming the lowest stretch of Ganga- Bhagirathi-Hooghly Riverine System. The lower Hooghly reach between Kalyani to Birlapur, is within the zone of tidal influence and is about 100 river km in length. The region falls under the Calcutta Metropolitan District (CMD) and is also highly industrialized and densely populated.

The annual mean range of tidal variation at Garden Reach, has been reported to be about 2.1 m and 4.2 m during the neap and spring tides, respectively. The river within the region is of meandering nature and variable cross-section. The river width varies between 450 meters (at Rabindra Setu) to about 1100 meters (at Birlapur) at the mean tide level. At many place within the study zone river depth is about 8-9 meters during the mean tide level.

Hooghly serves as the major source of fresh water to a host of townships on its both banks. Within the CMD area of about 1368 square kilometers, river Hooghly provides the main source for water supply to 2 municipal corporations, 23 municipalities, 37 non municipal towns and 544 rural mauzas with a total population of over 10 million. The river Hooghly also serves as an important navigational system and provides the natural drainage for the wastewater generated in the region. In addition, a large number of industries also draw water for their usage from the river. The developmental activities of the region are, thus, inextricably linked with the health of river Hooghly.

Inventory of Wastewater Discharge

Inventorisation of about 100 km tidally influenced stretch of river Hooghly between Kalyani and Birlapur and on Tolly's Nullah (about 20 km in length) was carried out by the NEERI in its study in 1993 (NEERI, 1993). These studies indicated that possibly more than 300 wastewater drains, having a wide variety of flow and characteristics, contribute to the pollution of the river Hoghly within the 100 km length.

Water Quality of River Hooghly at Selected Points

Since in a river water quality system, the pollutants form a dynamic equilibrium between the sediments and the water column, the analyses of bottom sediments provide important observations on the true water quality status of such a system.

The worst case river water quality estimation in summer indicates that the region receives immense quantity of waste loads from the urban areas without any treatment. Following sections present the status of the water quality at various locations in the river Hooghly's tidally influenced stretch.

The general water quality survey at selected points in river Hooghly at various locations was conducted during the summer of 1991 (Figure 20.1). The results of this survey in the form of physicochemical , bacteriological, trace metals and biological characteristics of Hooghly water quality are presented in Tables 20.1-20.7

Table 20.1 indicates that though the values for various physicochemical parameters some times show significant differences for high tide and low tide observations, no clear uniform pattern for all sampling points is discernible for any of the parameters. Similar observations are made on the across the stream variations. It may, therefore, be concluded that the local effects such as the proximity of a wastewater discharge, mixing pattern and local hydrodynamics and bottom deposits play an important role on the water quality in the river.

The results of trace metals analyses is presented in Table 20.4. The observations indicate that the stretch is specifically contaminated in terms of Iron, mercury, Lead and Manganese. Levels of Cadmium and Chromium also are above the drinking water limits in some regions of the river stretch.

The bacteriological quality observations presented in Table 20.5, also indicate a high degree of contamination of river water from wastes of domestic and animal origin. As regard to the biological indicators of water quality, it is observed from Table 20.6 that the diversity index for zooplanktons varies from 0.91 to 2.78, whereas, the total count for phytoplankton is observed to vary between 300 to 1600 per 100 ml over the study region as indicated in Table 20.5 – 20.7. It is observed that though in some instances the zooplankton diversity and photoplankton count differs during the low and high tide periods, no clear common pattern among all sampling stations is discernible in this regard. Further, any significant change of species composition is also not observed in the low and high tide samples.

Table 20.1: Status of Water Quality at Selected Points in River Hooghly

(Physical Parameters)

Sampling Points	Water Quality Characteristics							
	Turbidity (NTU)		T.S (mg/l)		T.D.S. (mg/l)		Conductivity (umhos/ cm)	
	LT	HT	LT	HT	LT	HT	LT	HT
Chinsurah	125	110	348	302	176	193	251	275
Seorafuly	105	98	311	265	180	182	237	240
Nimtala	118	110	296	266	175	179	250	251
Budge Budge	145	95	304	241	201	165	287	276

LT= Low Tide, HT = High Tide

All results are of the samples collected from the middle of the river

Though phytoplankton observations show the presence of three major groups of planktors, namely, bacillariophyceae (diatom), chlorophyceae and cyanophyceae, the diatoms constitute more than 90 per cent of the total count at all sampling stations. In cognizance of low conductivity and chloride values, these observations indicate very mild organic pollution state in most part of the river stretch.

Table 20.2: Status of Water Quality at Selected Points in River Hooghly

(Chemical Parameters)

Sampling Points	Water Quality Characteristics											
	Total Alkalinity as $CaCO_3$		Total Hardness as $CaCO_3$		Ca-Hardness as $CaCO_3$		Mg - Hardness as $CaCO_3$		Chlorides as Cl		Nitrate as N	
	LT	HT	LT	HT	LT	HT	LT	HT	LT	HT	LT	HT
Chinsurah	136	136	126	134	72	84	54	50	11	11	0.45	0.64
Seorafuly	136	138	140	130	86	64	54	66	12	12	0.39	0.50
Nimtala	138	136	134	136	50	44	84	92	13	13	0.78	0.42
Budge Budge	140	140	136	144	116	84	20	60	18	16	0.40	0.45

* Units : mg/l ; LT= Low Tide, HT = High Tide

All results are of the samples collected from the middle of the river

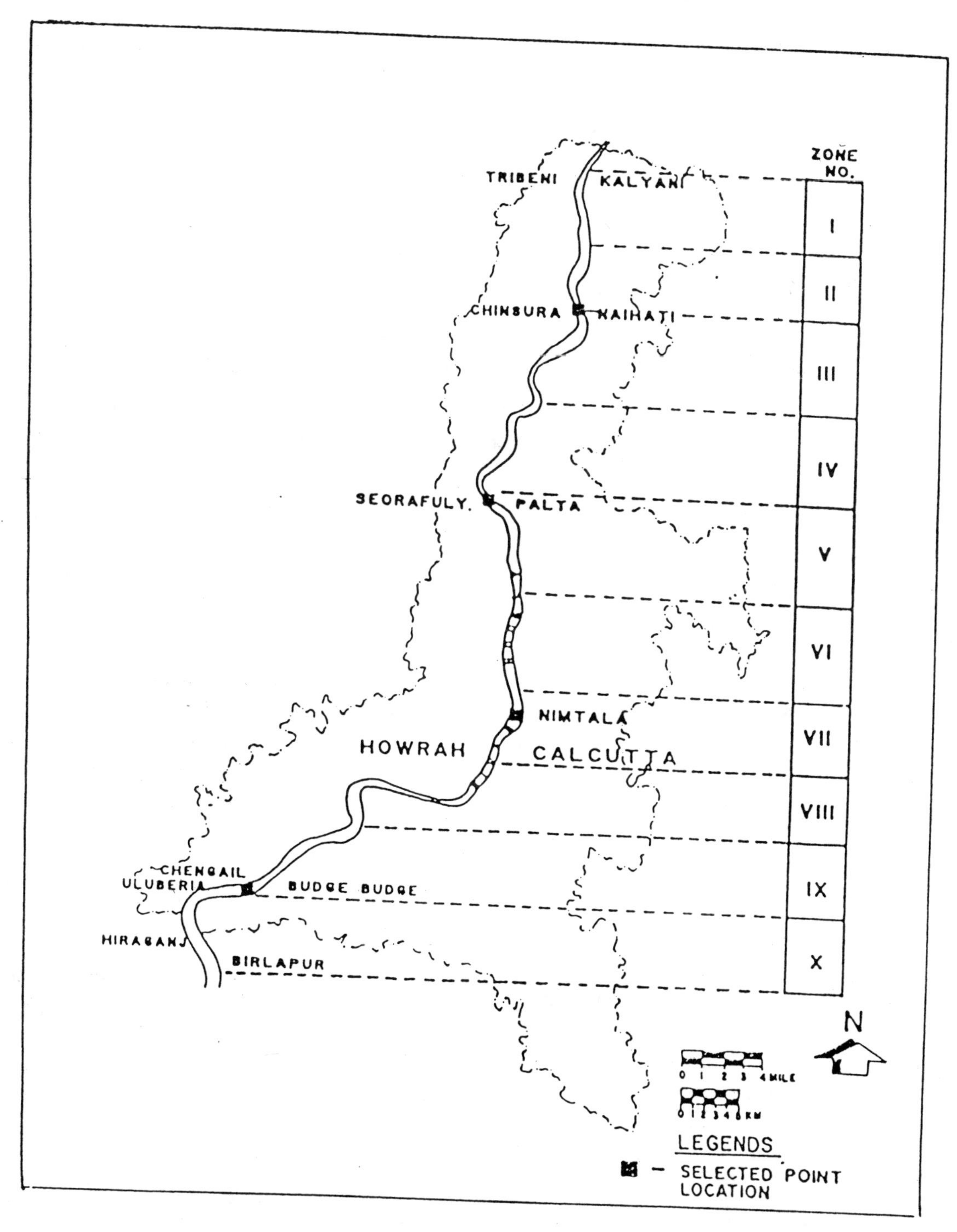

Figure 20.1: Sampling Locations for Water Quality at Selected Points in River Hooghly

The results of bottom sediment analysis in terms of total organic carbon content and metal concentration are presented in Table 20.8. The results show generally higher metal contamination of sediments in the upstream zones of the river in comparison to the downstream zones. Similar observation are made for the total carbon contents of the sediments.

Table 20.3: Status of Water Quality at Selected Points in River Hooghly

(Chemical and Nutrient Parameters)

Sampling Points	Water Quality Characteristics*									
	DO		BOD		COD		Total Phosphate -as P		Total Nitrogen -as N	
	LT	HT	LT	HT	LT	HT	LT	HT	LT	HT
Chinsurah	6.9	6.7	1.6	2.8	15	15	5.2	0.49	3.92	3.92
Seorafuly	7.8	6.8	1.5	1.1	11	11	0.11	0.08	5.32	4.48
Nimtala	6.0	5.9	1.3	1.9	8	8	0.56	0.18	6.44	5.04
Budge Budge	6.1	5.9	1.9	1.4	8	12	0.14	0.18	2.84	5.04

* Units : mg/l ; LT= Low Tide, HT = High Tide
All results are of the samples collected from the middle of the river

Table 20.4: Status of Water Quality at Selected Points in River Hooghly

(Metal Concentration)

Sampling Points	Water Quality Characteristics											
	Cadmium mg/l		Chromium mg/l		Lead mg/l		Zinc mg/l		Manganese mg/l		Mercury µg/l	
	LT	HT	LT	HT	LT	HT	LT	HT	LT	HT	LT	HT
Chinsurah	BDL	0.01	0.22	0.15	0.08	0.09	0.32	0.08	0.10	0.14	0.35	1.8
Seorafuly	0.01	0.01	0.03	0.06	0.06	0.06	0.21	0.59	0.14	0.32	1.3	0.6
Nimtala	0.01	0.01	0.02	0.01	0.11	0.05	0.10	0.11	0.12	0.28	0.8	0.8
Budge Budge	0.01	0.01	BDL	--	0.01	0.13	0.11	0.08	--	0.30	1.3	1.3

* Units : mg/l ; LT= Low Tide, HT = High Tide, All results are of the samples collected from the middle of the river; BDL -Below Detection Limit,
Detectable limit for the elements are :
Cadmium - 2µg/l ; Chromium- 10µg/l ; Lead-10µg/l
Zinc- 10µg/l ; Manganess- 10µg/l ; Mercury-0.5 µg/l

Table 20.5: Status of Water Quality at Selected Points in River Hooghly

(Bacteriological Parameters)*

Sampling Points	Water Quality Characteristics					
	Fecal Coliforms		E.Coli		Fecal Streptococci	
	LT	HT	LT	HT	LT	HT
Chinsurah	2.3×10^3	9.3×10^3	9.1×10^2	4.3×10^3	9.1×10^2	2.3×10^3
Seorafuly	4.3×10^3	9.3×10^3	4.3×10^3	9.3×10^3	2.3×10^3	2.3×10^3
Nimtala	4.3×10^3	9.3×10^4	2.3×10^3	4.3×10^4	2.3×10^3	4.3×10^4
Budge Budge	9.3×10^3	7.5×10^3	9.3×10^3	3.6×10^2	2.3×10^3	4.3×10^3

LT= Low Tide, HT = High Tide

* Units : MPN/l00ml

All results are of the samples collected from the middle of the river

Table 20.6: Status of Water Quality at Selected Points in River Hooghly

(Bacteriological Parameters)*

Sampling Points	Water Quality Characteristics (Percent Composition and Index Values of Zooplankton)											
			% Zooplankton in Groups									
	Total Number per m^3		Protozoa		Rotifera		Cladocera		Copepoda		Shannon Weaver Diversity Index	
	LT	HT	LT	HT	LT	HT	LT	HT	LT	HT	LT	HT
Chinsurah	4000	4000	33.33	50.00	--	16.67	--	--	66.67	33.33	1.46	2.28
Seorafuly	2800	4000	--	20.00	14.29	30.00	14.29	--	71.42	50.00	2.13	1.97
Nimtala	6000	4000	--	--	5.00	30.00	7.50	--	87.50	70.00	1.77	2.24
Budge Budge	35000	16000	28.57	--	2.86	12.50	--	--	68.57	87.50	2.18	1.77

LT= Low Tide, HT = High Tide

All results are of the samples collected from the middle of the river

Table 20.7: Status of Water Quality at Selected Points in River Hooghly

(Bacteriological Parameters, Phytoplankton)*

Sampling Points	Water Quality Characteristics	
	Phyto-plankton counts /100 ml	
	LT	HT
Chinsurah	1200	1300
Seorafuly	1600	500
Nimtala	700	1200
Budge Budge	1000	300

LT= Low Tide, HT = High Tide
All results are of the samples collected from the middle of the river

Table 20.8: Results of Bottom Sludge Analysis at Selected Points of River Hooghly

Sampling Point Location	Total Organic Carbon (%)	Metal Concentration (μg/g)				
		Cadmium	Chromium	Lead	Zinc	Manganese
Chinsurah (Left Bank)	2.56	1.0	29	27	55	600
Seorafuly (Left Bank)	1.84	2.0	24	20	94	600
Nimtala (Left Bank)	1.60	1.0	24	22	53	500
Budge Budge (Left Bank)	1.10	1.0	21	11	39	500
Budge Budge (Right Bank)	1.31	1.0	20	23	42	500

N.B. = Samples collected during lowest low tide period

The complete length of river stretch with CMD area is tidally influenced and therefore, is subjected to very dynamic currents and tidal changes. It has been observed that at the time of low as well as high tide, strong down stream and up stream currents developed in the region. It is likely that due to the turbulence and movement of large water mass over large distances with the tide, the pollutants released in the river get dispersed over a wide region.

Because of such dispersion and large quantum of dilution available even during the lean flow season, in spite of significant inputs of domestic waste in certain areas the physicochemical parameters even in these zones do not show a significant elevation.

The discharges in the river finally meets then coastal areas and the effect on the estuarine and marine ecosystem is significant.

CONCLUSIONS

The water quality survey indicates that wastewater discharged into the river gets distributed over fairly long stretches. For effective surveillance monitoring after the implementation of the Ganga Action Plan, observations within the mixing zones for major drain during the low and high tides shall be necessary. It is desirable, therefore, that the regions of such zones are identified and documented for all major drains prior to their diversion for delineation of efficient river monitoring network in future. Use of appropriate mathematical models will be of great use to facilitate optimum use of resources in this respect.

The effects of pollutants ultimately gets manifested in the biological components of water quality system. Benthos, in this respect is a potential to provide a good indication of river water quality status in the river. Regular monitoring of Benthos in the critical zones is therefore, recommended to be made on integral part of future monitoring in the river.

REFERENCES

Alder, C. A. (1973): Ecological Fantasies, , Green Eagle Press, 1973

CPCB (1993): Monitoring of Indian Coastal Waters, Summary report, Central Pollution Control Board, New Delhi, 199.

GOI, (1974): Water (Prevention and Control of Pollution) Act, 1974 Section 2(I), Government of India.

NEERI (1993): Rapid Estimation of Point Sources Discharging into River Hooghly within CMD Area, Report by National Environmental Engineering Research Institute, Nagpur, 1993

Salas Henry T. (1994): Submarine Outfalls A viable Alternative for Sewage Discharges of Coastal Cities in Latin America and Caribbean, Report by Henry T. Salas for Pan American Centre for Sanitary Engineering and Environmental Sciences, 1994